AF594945

Business Models and Sustainable Development Goals

Business Models and Sustainable Development Goals

Editor

Prescott C. Ensign

MDPI • Basel • Beijing • Wuhan • Barcelona • Belgrade • Manchester • Tokyo • Cluj • Tianjin

Editor
Prescott C. Ensign
Wilfrid Laurier University
Canada

Editorial Office
MDPI
St. Alban-Anlage 66
4052 Basel, Switzerland

This is a reprint of articles from the Special Issue published online in the open access journal *Sustainability* (ISSN 2071-1050) (available at: https://www.mdpi.com/journal/sustainability/special_issues/sustainable_goals_business_models).

For citation purposes, cite each article independently as indicated on the article page online and as indicated below:

LastName, A.A.; LastName, B.B.; LastName, C.C. Article Title. *Journal Name* **Year**, *Volume Number*, Page Range.

ISBN 978-3-0365-4305-5 (Hbk)
ISBN 978-3-0365-4306-2 (PDF)

Cover image courtesy of Prescott C. Ensign

Contents

About the Editor

Prescott C. Ensign

Prescott C. Ensign is a Professor at the Lazaridis School of Business and Economics, Wilfrid Laurier University, Waterloo, Ontario, Canada, and holds a PhD in Management (Université de Montréal/École des Hautes Études Commerciales). His research focus is in the areas of global business and strategy; entrepreneurship and innovation; economic development at the periphery and socio-political aspects of indignity; and business model innovation, sustainability, and social enterprises.

Editorial

Business Models and Sustainable Development Goals

Prescott C. Ensign

Lazaridis School of Business & Economics, Wilfrid Laurier University, Waterloo, ON N2L 3C7, Canada; ensign@wlu.ca

In 2015, all 193 member countries of the United Nations adopted the 2030 Agenda for Sustainable Development. It includes 17 Sustainable Development Goals (SDGs). Building on the principle of "leaving no one behind," it emphasizes a holistic approach to achieving sustainable development [1]. The 2020 environmental, social and governance (ESG) scoring and reporting document from the Organization for Economic Co-operation and Development (OECD) notes that sustainability investing has grown, primarily due to the number of funds and investors that have added ESG approaches to their overall agenda. Corporations, central banks and the public sector are placing a new emphasis on a greener environment and low-carbon economy [2]. The 2020s was to be a decade of action but progress has been slow, stalled or reversed in meeting the 17 SDG targets [3]. OECD's quantitative analysis provides an indication of the progress made and challenges still ahead with regard to sustainable development. The wide variety of metrics, methodologies, and approaches indicate a high number of disparate outcomes that are open to interpretation [4].

The outcome of the 26th UN Climate Change Conference (COP26) held in Glasgow Scotland (November 2021) is also open to interpretation. COP26 participants took steps to address the climate crisis by agreeing "to revisit and strengthen the 2030 targets in their nationally determined contributions ... by the end of 2022" [4]. These steps fall far short of delivering the national commitments necessary for a unified global effort to limit annual planet warming to 1.5 °C [5,6]. To find real SDG and ESG progress, we have to look past "agreements" by world leaders and metrics designed to influence institutional investors. We need to identify, explore and examine SDGs at multiple organizational levels—firm, sector, regional and national levels. This is where change, success and hope for our planet's future rests.

This Special Issue on "Business Models and Sustainable Development Goals" presents five research studies that examine transformative business models designed to support achievable sustainable development. Every organization—from start-ups, small and medium size enterprises, multinationals, social enterprises, hybrids, cooperatives, non-profits to NGOs and government agencies—has a business model. It reflects management's explicit or implicit hypotheses on why and how the organization functions. The business model is the engine that powers an organization, defining the value proposition of the venture, how it balances resources with the ecosystem where it operates, how it generates cash flow and creates value [7]. Changes to an organization's business model are recognized as a fundamental approach to implementing innovations for sustainability. The capability to transform or transition to new business models is an important source of sustainable competitive advantage and provides leverage to improve the performance of organizations. Borrowing from Geissdoerfer and Vladimirova, a sustainable business model (SBM) includes pro-active management, monetary and non-monetary value for a broad range of stakeholders, and a long-term perspective [8].

The first study by Ramanauskaitė, The Role of Incumbent Actors in Sustainability Transitions: A Case of LITHUANIA, explores actions at the organization, sector, regional and national level by five influence-shaping organizations that espouse a sustainability orientation. As a country, Lithuania has faced precipitous change since it obtained independence from the Soviet Union and is now an EU and NATO member. This has been a

Citation: Ensign, P.C. Business Models and Sustainable Development Goals. *Sustainability* **2022**, *14*, 2558. https://doi.org/10.3390/su14052558

Received: 11 February 2022
Accepted: 16 February 2022
Published: 23 February 2022

time of accelerated transition for companies moving from a planned economy to a free market socio-economic system. The paper examines the next question: what will the transition to sustainability be like? An extensive review of the relevant literature suggests that resistance by key incumbent actors will negatively affect sustainability at multiple levels. However, interviews and secondary data obtained in the study reveal the following. The five powerful and prominent companies that are making transitions to sustainability were basically promoters and sponsors but still face tension and resistance. Ramanauskaitė provides insight into the need for researchers and practitioners to understand the internal and external issues involved in transitioning a business model compared to transforming a business model. The study also provides insight on understanding the impact of corporate leadership on sustainability—insight that goes beyond the firm and sector to one that focuses at the national level.

The second paper by Ensign, Roy and Brzustowski, Decisions by Key Office Building Stakeholders to Build or Retrofit Green in Toronto's Urban Core, moves the focus of sustainability and business model inquiry to the urban office building sector and regional metroplex level. It focuses on aligning new and existing construction decisions to achieve sustainable environmental goals, e.g., the reduction of greenhouse gas (GHG) emissions. Motivation for the study was based on the following. The federal and provincial governments in Canada had set GHG emissions mitigation targets for 2030 based on the Paris Climate Agreement and the Pan-Canadian Framework. Toronto's City Council declared a climate emergency on 2 October 2019, joining a global call to recognize the urgency of the climate crisis and adopting a stronger emissions reduction target for Toronto—net zero by 2050 or sooner. In December 2020, Ottawa announced plans to raise the federal carbon tax from CAD 30/tonne to 170/tonne by 2030. Would these actions impact the office building sector? The aim of this study was to determine whether key stakeholders (real estate developers, institutional investors and owner landlords) would choose a business model for or against sustainability. The authors introduce a six-stage real estate development process approach used by the sector as a lens for identifying and understanding the point at which the decision for conventional or green construction is made. In urban core real estate development, the metrics of each project are unique and the process often more political than economic. Two hypotheses were proposed: Hypothesis (H^1) LEED (Leadership in Energy and Environmental Design) certification has a significant positive factor in the asking rent (market value); and Hypothesis (H^2) financial drivers will be more influential than non-financial drivers in a key stakeholder's decision to pursue LEED certification. Pearson correlation and linear regression analysis of data on 16 LEED and 52 conventional buildings did not find LEED certification to be statistically significant in explaining the variance in the net asking rent (market value). Interviews conducted with senior executives engaged in Toronto's real estate development sector provided qualitative data to test H^2. The expert informants did not view LEED certification as a primary deciding factor but as one of a combination of factors that impact a firm's financial bottom line. Transcripts of these interviews offered insight on positional differences between real estate developers, institutional investors and commercial office space renters on SDGs, ESG and GHG emissions. Even when LEED construction is selected, the values, goals and actions of the key stakeholders involved represent a diversity of business models.

The third paper by Geldres-Weiss, Gambetta, Massa and Geldres-Weiss, Materiality Matrix Use in Aligning and Determining a Firm's Sustainable Business Model Archetype and Triple Bottom Line Impact on Stakeholders, presents a tool that can be used to enhance or transform a company's traditional business model (TBM) to a sustainable business model (SBM). It introduces the materiality matrix (MM) as a way for a company to review its business model with a more comprehensive understanding of the value creation processes that maximize the total value, not just the financial value for stakeholders. By way of explanation, the total value for stakeholders is a multidimensional view of material issues in environmental, social and governance/economics (ESG) as well as corporate social responsibility (CSR) that collectively influence the value creation process. Data used to test the MM design include what are typically presented during GRI (Global Reporting Initiative) sustainability assessment preparation. Linking the MM to the business model provides the ability to gauge, measure and account for value creation from both an internal and external perspective. MM can create a triple bottom line impact through shaping strategic elements in the firm's business model. The study looked at transforming the company's existing TBM into a more sustainable one (inside-out approach) by enabling the identification of the most appropriate business model archetype to incorporate innovation into its SBM (outside-in approach). A case study is presented of the large globally prominent Chilean winery Viña Concha y Toro S.A. (VCT) to illustrate how MM can be used in transforming TBM into an SBM archetype. Based on GRI reports and interviews the authors chart and analyze the changes in VCT's material issues and associated prioritization between 2017 and 2019. The VCT case facilitates understanding how MM-aided changes impact an SBM archetype and value creation process. The case study also reports on the transformational alignment of the MM with implementation of a triple layered business model canvas (TLBMC) and SBM archetype. VCT's SBM holds that economic success goes hand in hand with caring for the environment, making rational use of natural resources and a commitment to people and the social sphere where VCT operates during each stage of the value chain.

The fourth paper by Mansell, Philbin and Broyd, Development of a New Business Model to Measure Organizational and Project-Level SDG Impact—Case Study of a Water Utility Company, introduces a methodological approach for linking local project-level sustainability performance to global SDG targets. Infrastructure construction project success has traditionally been measured (time, cost and scope/quality) when a project is completed (delivered). If engineers and project managers are the ones responsible for a construction project's sustainability, then performance measures agreed on by all the stakeholders need to begin at the project design phase. A recent Institution of Civil Engineers' survey of practitioners cites four critical factors for measuring a project's SDG success: clear definitions, holistic performance measurement tools, alignment of business priorities, and strong leadership. Having defined different ways of classifying project success, the authors suggest a new SDG transformation business model process for infrastructure projects and the infrastructure construction sector in general—the SDG infrastructure Impact-Value Chain (IVC). The conceptual basis embedded in the IVC is that there is a "golden thread" that links tactical SDG success during project construction and delivery to strategic SDG success that is embodied in longer-term post-project outcomes. Based on interviews and secondary sources, a case study of Anglian Water is presented to demonstrate the IVC new business model approach's use at the project design phase to align stakeholders on why/when/how/what SDG targets to measure. Anglian Water is one of the UK's largest water utility companies and a leading triple-bottom-line (TBL) sustainability supporter with a vision to create a resilient environment that allowed sustainable growth and the ability to cope with the pressures of climate change. Anglian Water's Wisbech project was then chosen as part of this case study because data on its delivery are open source on the Internet. This project is an IVC example because it was delivered by Anglian Water and its partners as part of their commitment to make a long-term impact on the market town of Wisbech (for more than the five years that the initial infrastructure project covered). Anglian Water wanted to assess whether a broad programme of social, economic and environmental changes in the lives of those in local communities can be linked to using the IVC. The SDG targets offer a framework to address the more diffuse outcomes and impacts that might not have been defined and measured using traditional project measurement approaches. The results of the case study investigation have indicated that there is a verifiable link across the IVC of activities–inputs–outputs during the "in-project" phase, connecting to the "post-project" outcomes and SDG impacts. The practical application of IVC is significant. With improved linkage of tactical delivery to strategic SDG impacts, improved investment decisions will be made and systemic-level lessons can be applied to increase the likelihood of success in achieving the SDG 2030 targets.

The fifth paper by Doroteja, Marolt and Pucihar, Information Technology for Business Sustainability: A Literature Review with Automated Content Analysis, addresses the issue of managing sustainability in a digitalized environment. The authors conducted a systematic literature review and content analysis of 61 articles published between 2008 and 2020, providing insight on the role of information technologies (IT) in sustainable business models. They found that the majority of the research in this rapidly developing interdisciplinary field was from the European Union. This is not surprising since the 2018 European Commission states that competitiveness in the coming years will depend on sustainability and the ability to exploit IT. This strong interest in sustainability issues by EU policy makers and enterprises raises an important question. Could this be related to the two-tier system of corporate governance used in most EU enterprises? This dual system—a structure of management and a supervisory board with different roles—creates opportunities to focus on different values (e.g., environmental, social and governance criteria). Decisions in corporations with a unitary board system that consists of a single board of directors and absentee shareholders such as in the USA tend to focus on economic gains rather than social and environmental concerns. During the last decade, digital transformation has resulted from the emergence of new technologies (e.g., social, mobile, analytics, artificial intelligence, cloud, high-performance computing, Internet of Things, and robotics) that have an impact on individuals, enterprises, organizations and society. In a business context, digital transformation often refers to a process of redesign or innovation of a business model from the adoption and use of IT to create digital capabilities. This study suggests that the transformation of IT's role is bringing about sustainability. IT and digital capacities are not limited to reactive innovation and application. IT can play a major role as an enabler and driver for pro-active transformation of a traditional business model to a sustainable business model as well as create new value propositions.

The goal of this Special Issue was to provide a platform for scholars to share theory-based research, conceptualization, and case studies. These papers broaden and accelerate our understanding of innovative business models that support sustainable development. I trust that you the reader will find that this eclectic collection of papers presents an interesting examination of the internal and external challenges of aligning business models to achieve SDGs. A sincere thank you to all the Special Issue authors for their contribution to enhancing our understanding and providing valuable insights. A special thank you to the external peer reviewers for providing valuable feedback, comments and suggestions to improve the significance of the contributions. Finally, we want to express our thanks to the staff members at the MDPI editorial office, in particular Dana Shen, for their valuable support and encouragement.

Funding: Ensign received support from the Social Sciences and Humanities Research Council of Canada as well as a Principles for Responsible Management Education seed grant from the Lazaridis School of Business & Economics.

Conflicts of Interest: The author declares no conflict of interest.

References

1. United Nations. The 17 Goals. Available online: https://sdgs.un.org/goals (accessed on 23 January 2022).
2. Boffo, R.; Patalano, R. *ESG Investing: Practices, Progress and Challenges*; OECD: Paris, France, 2020; Available online: https://www.oecd.org/finance/ESG-Investing-Practices-Progress-Challenges.pdf (accessed on 23 January 2022).
3. Stockholm Environmental Institute. Let's Get the SDGs Back on Track. 2020. Available online: https://www.sei.org/perspectives/lets-get-the-sdgs-back-on-track/ (accessed on 23 January 2022).
4. The Law Society. Reflecting on COP26: What Were the Key Outcomes? 2021. Available online: https://www.lawsociety.org.uk/topics/climate-change/reflecting-on-cop26-what-were-the-key-outcomes (accessed on 23 January 2022).
5. University of Edinburgh. Was COP26 a Success? Available online: https://www.ed.ac.uk/impact/opinion/was-cop26-a-success (accessed on 23 January 2022).
6. Bomberg, E. COP26: Are we Looking for Success in the Wrong Places? University of Edinburgh. Available online: https://www.ed.ac.uk/impact/opinion/cop26-looking-for-success-in-all-the-wrong-places (accessed on 23 January 2022).

7. Teece, D.J. Business models, business strategy and innovation. *Long Range Plan.* **2010**, *43*, 172–194. [CrossRef]
8. Geissdoerfer, M.; Vladimirova, D.; Evans, S. Sustainable business model innovation: A review. *J. Clean. Prod.* **2018**, *198*, 401–416. [CrossRef]

MDPI

Article

The Role of Incumbent Actors in Sustainability Transitions: A Case of LITHUANIA

Joana Ramanauskaitė

School of Economics and Business, Kaunas University of Technology, LT-44249 Kaunas, Lithuania; joana.ramanauskaite@ktu.edu; Tel.: +370-630-83-955

Abstract: To explore what roles incumbent actors take in sustainability transitions, this paper investigates the current situation in the scientific literature, which reveals a shift from opponents to promoters and the case of a post-Soviet transitioning economy that is exemplified by examining five sustainability-oriented incumbent actors in Lithuania. A single case study design is selected as a methodological approach, illustrated by empirical data from interviews and secondary sources (corporate websites and sustainability reports). These examples provide insights on the initiatives the organisations that are already interested in sustainability take to promote sustainability ideas and be active members of the transition themselves, supporting the contemporary view of incumbents as agents of sustainable transitions.

Keywords: incumbents; sustainability transitions; sustainability; transitioning economy; case of Lithuania

Citation: Ramanauskaitė, J. The Role of Incumbent Actors in Sustainability Transitions: A Case of LITHUANIA. *Sustainability* **2021**, *13*, 12877. https://doi.org/10.3390/su132212877

Academic Editor: Prescott C. Ensign

Received: 19 October 2021
Accepted: 18 November 2021
Published: 21 November 2021

Publisher's Note: MDPI stays neutral with regard to jurisdictional claims in published maps and institutional affiliations.

1. Introduction

The role of business and organisations in sustainability transitions is receiving increased attention from scholars in the field [1]. Sustainability transition could be explained by examples in the current trends for the promotion of high value-added sustainable products, alternative and renewable energy sources, and circular or bio-economies [2–4]. The European Union (EU) recognises the importance of sustainability issues and includes them in its policy agenda aimed at building a knowledge-based bio-economy [3,5]. Sustainability transitions refer to purposeful, long-term, multidimensional, fundamental transformations of socio-technical systems towards more sustainable modes of production and consumption, requiring participation of different types of actors [6–8] and therefore, certain structures of analysis. Multi-level perspective (MLP) could theoretically be one of the main mechanisms to explain these processes [2,6,9].

To understand the sustainability transitions, a holistic picture is necessary. The MLP framework provides a deconstruction of sustainability transitions into three levels: niche, regime, and landscape [3,10,11], although it is sometimes considered too simple, relatively straightforward, and unable to capture the inherent complexity of the system change [12–14]. It is interesting and complementing to look at the organisational level of the MLP, best described by the regime level. A regime can be defined as a group of actors sharing a set of rules that are unique to that regime. Interconnectedness and partial overlapping of different regimes guiding actors in a socio-technical system is referred to as the socio-technical regime [2]. There are different actors at play when transitioning towards sustainability, because sustainable development does not address the needs of an exclusive group, but rather incorporates the interests of multiple groups of social actors and even of different generations [15]. Organisations are an important part of sustainability transitions, transforming societies and markets; therefore, the business role in this transition is an important aspect to be explored, complementing the outlook regarding the type and size of the organisations [16–18]. Socio-technical regimes that have been long established may include large, influential organisations at play. These established prominent actors with a lifelong history are large both in terms of personnel and revenue, have political power, but are often black-boxed and labelled as homogeneous industry structures with predefined roles and relations, referred to as "incumbent" in the scientific literature [19,20]. Interestingly, incumbent actors, even big players in industry and overall country or regional development, were overlooked until quite recently (i.e., [4,20–22]) and are being started to be investigated by researchers of sustainability transitions.

The role of incumbent actors is still being discussed; it is beneficial to look not only at the developed countries, but also see how these possible game changers might affect the developing economies or economies-in-transition. It is noted that the current literature lacks insights from a geographical perspective in the overall context [23]. Research from Eastern European post-Soviet countries is left at the border of peer-reviewed scientific journals, with probable lower acceptance rates of submissions [24]. However, the context of post-Soviet countries is an interesting area for sustainability transition research, mainly because of the recent and ongoing transition from a socialist planned to capitalist market economy [25,26]. There was a shift not only in economical arrangements, but also in social and psychological attributes. The past is not easily forgotten, and some of the inhabitants of post-Soviet countries may feel reluctant to engage in sustainability due to the similarity between the communist and sustainability ideas, i.e., shared goods, renting instead of buying; even some nomenclature might resemble the days under the "regime" (i.e., corporate "social" responsibility, "socioeconomic", etc.). However, this paper is not aimed at studying the psychology behind the willingness to participate in the sustainability movement. These remarks on the possible hesitation of post-Soviet communities to address the sustainability issues might be among the reasons why this context is interesting and should be addressed as part of a worldwide research concern.

It is important to study how businesses contribute to sustainable development. This paper relates to Targets 8.3, 12.6, and 12.8 of the Sustainable Development Goals (SDGs) encouraging enterprises, especially influential ones, such as incumbent actors, to actively promote sustainable development, build awareness, and participate in policy and decision-making, creating a more sustainable environment [1]. This paper looks at how incumbent organisations promote sustainable development, encourage their stakeholders to be more sustainability-oriented, and how they act themselves, thus accelerating the progress towards SDGs in the wake of the crises of the modern world [1].

This paper takes an inductive research approach. The idea for this paper came while investigating contradictions and tensions in corporate sustainability transitions, which revealed some interesting examples of incumbent actors in the transitioning economy of Lithuania. As a member of the European Union and one of the post-Soviet countries, Lithuania might expose some insights on sustainability transitions and the role of incumbent actors in them. This paper is structured as follows. Firstly, the theory is presented for perceptions on incumbent actors in sustainability transitions and the context of post-Soviet transitioning economy. Secondly, the methodology and a sample of incumbent actors in the context are described. Following this, the research results are presented. Lastly, conclusions and discussions are drawn from the observations in the paper.

The research objective of this paper is to explore the initiatives of incumbent actors in sustainability transitions in a transitioning economy. The research question is formulated as follows: how do the incumbent actors in transitioning economy contribute to sustainability transitions?

2. Theoretical Background

2.1. Role of Incumbent Actors in Sustainability Transitions

As mentioned previously and agreed upon by sustainability transition researchers, the role of incumbent actors in sustainability transitions is underexplored and is a topic for future research. As was discovered during the scientific literature review, several studies explored the definition of an incumbent actor in sustainability transitions. Their definitions are provided in Table 1 below.

Table 1. Definitions of incumbent actor in the scientific literature on sustainability transitions.

Source	Definition	Keywords
Apajalahti [19]	"By 'incumbent' most of the studies I have come across specify this to be an organisation or an actor with a longstanding history; it is large in size, both in terms of personnel and revenue; in most cases it is a well-known firm; and it has political and economic power (there are also other actors, such as industry associations, that hold power in the sense that they represent the majority of an industry)." (pp. 31–32)	Longstanding history Large size High income Well-known Power
Borghei [20]	"However, studies in this field have largely neglected established actors (i.e., incumbents) and their potential role(s) in facilitating societal transformations. Incumbents are black-boxed as a homogenous set of actors with a pre-defined role in relation to the established industry structures (Geels, 2002; Holtz et al., 2008)." (p. 4, citing [11,27])	Established Potentially powerful Homogenous Inveterate
Sovacool et al. [28]	"Drawing from Johnstone et al. < . . . >, we define incumbent actors or institutions as those "that often have vested interests in maintaining the status quo rather than enabling transitions and will often act to strategically protect their privileged position" within a given socio-technological regime." (p. 3, citing [29])	Status-quo Privileged Protective
Hengelaar [30]	"Innovation literature commonly defines incumbents based on their established nature: "firms that manufactured and sold products belonging to the product generation that preceded the radical product innovation" (Chandy, Tellis 2000, Henderson 1993, Mitchell 1991, Mitchell, Singh 1993)." (p. 18, citing [31–34])	Pre-radical innovation products Manufacturers
Kungl [35]	"Incumbents "are those actors who wield disproportionate influence within a field and whose interests and views tend to be heavily reflected in the dominant organization of the strategic action field" ([4]: p. 13)." (p. 14, citing [36])	Disproportionate influence Dominance Strategic field Field-bound
Smink, Hekkert and Negro [37]	"[I] ncumbents: the firms that mainly have competencies related to the current technological regime, and that (financially) benefit from existing practices." (p. 87)	Inveterate

These prominent, powerful organisations have a role in sustainability transitions and general change in local or regional environment development. However, how these actors shape the landscape, regime, or niche is still being debated. Earlier works assign a sort of restraining attributes, seeing incumbent actors as keepers of steady, unmovable regimes to stay the way they are [28]. More current studies note other traits of these prominent organisations, characterising them as promoters, sponsors of sustainable transition, those that can notice the change in the market and the needs of stakeholders and are able to respond to the most current trends and demands. Nevertheless, incumbency can be attributed not only to monopolies of firms or powerful governmental players, but also to varieties of different societal dominions, their levels and degrees (market, state, society, etc.) [38], that are not necessarily associated with large enterprises [17]. Incumbents pose a category of specific interest due to their potential power to steer change and entrusted interest [30]. The extracted keywords in Table 1 complement these notions, where authors mention the status-quo incumbents take [20,28,37], their ability to apply power [19,20,35]; however, the size or type of the organisation is not mentioned as often [19,30]. Research by van Mossel et al. [2] provided a categorisation of incumbents by their behaviour during transition: first to enter, follow into, remain inert, and delay the transition; also supported by other authors [3]. The exploration of the role of incumbent actors as perceived by scientific studies is described in the following paragraphs.

Van Mossel et al. [2] studied the behaviour of incumbent firms during transitions. They began with the notion that incumbents were mostly perceived as locked in their behavioural patterns and reacted passively to the social environment changes. Agreeing with this notion, Hörisch [39], Lindberg, Markard, and Andersen [40], Schaltegger et al. [17], and Turnheim and Sovacool [38] pointed out that incumbents were more likely to invest into economically sound innovations at the niche level, but not radical innovations that disrupt the regime, which the incumbents were more likely to preserve. The size of the company and bureaucratic inertia were among the reasons preventing investment in emerging innovations [4] and stalling sustainability transitions [2]. Sometimes incumbent actors show signs of persistent resistance to or slowing down of the transition because their business models are challenged, and their capabilities and resources are at risk [35,37,40,41], additionally blocking grassroot initiatives [3]. Therefore, a frequent hypothesis is that radical innovations are more likely to emerge from new entrants or smaller enterprises than incumbent actors [28], who commit to existing socio-technical regimes and oppose sustainability transitions [42] directing their resources accordingly.

The incumbent actors have access to excessive resources and might distribute them accordingly to their advantage. These resources (money, capabilities, competencies, influence) are built around the existing processes and routines, resulting in a lack of interest to invest in emerging technologies and breakthroughs that could disrupt the current status of incumbent organisations [4,38]. Changing the direction of resource investment would also require changes in the business model, viewed by the incumbent actors with reluctance due to the profitable core activities of economies of scale [4] and the current reputational status. Noticeably, incumbents are not expected to deploy their resources for social collective prosperity and distributional fairness [38]. Large incumbent firms are immersed in multiple regimes, which are also often incumbent, and the resources might be suspended there [2]. Therefore, resources of large incumbents might be a source of power that might be used for various reasons, not only tilting the areas that sustainability transitions struggle in [38], but also to assert the advantage politically.

Although incumbents respond to transitions, the response is not necessarily beneficial for further successful transition [2]; therefore, it is important to understand how incumbents shape their environment [40]. One of the ways incumbents can influence the sustainability transitions is to gain access to policymaking. Some sectors, i.e., energy and agriculture, are particularly politically powerful and might successfully resist fundamental change [42–44]. Possibilities for the incumbents to resist transitions might emerge in forming political coalitions, networks, or alliances [2,43]. By doing so, incumbents can strategically set technical standards and shape expectations and visions for stricter regulations [37], alter and manipulate information and knowledge [38,45]. Incumbent actors tend to favour their interest by actively shaping public policies [40,46] that might provide them with competitive advantage [38]. In any case, active response to sustainability transitions does result in changes in the behaviours of incumbent actors. One way is to resist and try to diminish the impact of ongoing change on their daily routines. Another solution would be to alter their own interorganisational habits, transforming their vision, reorienting their pathways, and gaining benefits by destabilising the existing regimes [38,47].

However, the perception of incumbents as resisters to sustainability transitions has been challenged. Although empirical insights on the role incumbent actors take on in sustainability transitions are scarce [39], incumbents are observed to contribute to the destabilisation or fragmentation of the regime [38,48]. The Schumpeterian look on incumbents notes that they are known for improving continuously and incrementally, whereas radical change is attributed to small new firms and entrepreneurs [30,49]; incumbents do contribute to developing innovative transitional technologies [50,51]. Newcomers to the sector could be perceived as challenging not only the market share of incumbents, but also their business models [41]. Although incumbents more often might be the keepers of existing regimes, some cases show their willingness to cooperate and invest in the strategic reorientation [41,52]. The incumbents collaborate with new entrants (with or without the intermediation), create common value, and enable newcomers, given their remarkable financial, human, and other resources [38,47,51]. This could considerably accelerate and direct transitions towards sustainability. Although incumbents are more likely to participate in reconfigurational changes, they might bring up and adopt economically optimal innovations to the regime from the niche level [39]. Precariously, incumbents tend to hold the significant power of what innovations to pick out and institute at the regime level [39]. However, there are studies showing that incumbents use their capabilities to develop several technological alternatives concurrently to change the directions of stable pathways [20,50]. Overall, there is contemporary research that shows a significant contribution of incumbents to sustainable transitions as proponents and promoters of new technologies and change [39,50,53].

Irrespective of the way the incumbents choose to participate in sustainability transitions, there are reasons behind those decisions. As analysed earlier, there are different behavioural patterns the incumbents might display. As the majority of the scientific literature suggests, incumbent actors show a tendency to be hesitant in taking an active role in developing radical innovations and technologies [54]. However, there are internal and external barriers or restraints for incumbents to participate in sustainability transitions. Common external impediments include weak institutions that lack directionality in shaping the transitional context, verbalisation of user preferences, straightforward policies, and regulations [28]. The interorganisational barriers are somewhat harder to define and require a more thorough consideration. Managers of incumbent firms may fail to adequately recognise disruptive threats [28] due to resistance to experimentation and transformation of their stable business processes and models [13]. Lack of knowledge of innovative products required by the market [4] might be at the core behind the fear of change. Established networks that support stable routines of organisations are often supported by technologies and developed competences that are threatened by the emerging new and radical innovations threatening to destabilise the current position of incumbents [20,28,54,55]. This notwithstanding, some of the incumbents with active resistance strategies focusing on niche innovations might sustain the carrying out of business as usual [56] for a while, but eventually, locking themselves into routines, technical capabilities and failure to recognise the destabilisation caused by disruptive technologies might result in the downfall of incumbent actors [57].

However, there are stimulating factors, promoting and encouraging incumbents to take an active and positive turn towards sustainability transitions that are being discussed in the current literature. Firstly, there are economic incentives at play that bring attractiveness and financial opportunities to push the transition process towards green niche innovations [42], whose policies have become more favourable for incumbents in the past few years, easing the transition [57]. Experimenting with new technological paradigms and niches provides impetus to overcome both internal and external barriers and allows to adapt to new markets faster, exploiting their current knowledge base and developing new competencies [4,17] and increasing their chances of survival, even though an incumbent might become dependent on the fate of the niche [2]. Introducing incumbents to sustainability ideas and societal missions might become important pushing these ideas into political level decision-making, simultaneously making the transition less disruptive, therefore, more attractive [28,40]. Moreover, incumbents with diverse capabilities and resource bases are more likely to survive in unstable environments [2]. Nonetheless, when faced with external pressure of new more sustainable entrants to the market, incumbents might engage in sustainability in their own way, achieving broader impact with the scale of the market they already share [22].

An outside stakeholder could bring a better vision and understanding of the current situation and of the benefits it could bring to the organisations. These external interested parties are "transition intermediaries—agents who connect diverse groups of actors involved in transitions processes and their skills, resources and expectation" [28] (p. 1). Sovacool et al. [28] dedicated a research to incumbent-oriented transition intermediaries and noted that this field was getting more prominent. Transition agents bring an upheaval and disruption to the regime, maintaining contact with stakeholders that have different interests in the transition. By becoming involved into these established ecosystems, transition intermediaries might cause conflict with incumbent business alliances [28,41]. The intermediaries are known for their push of new innovations from the niche level to regimes and protecting them from being overwhelmed by the resisting incumbents. This also requires intervention into policymaking. Providing acceptable prospects for both newcomers and incumbents, transition intermediaries might create win–win situations for the stakeholders, adding a third win for the sustainability transitions as a public good by using their abilities in either helping mobilise the resources, consulting on technology use, linking actors with similar interests and complementing resources, pushing environmental innovations and enabling the growth of niches, or lobbying at the political level [58]. Enabling new technologies offers a renewal for success that could be used for benchmarking by the incumbents and their regimes [41]. Collaboration is nonetheless key for enabling incumbents to invest in sustainability transitions.

Given the resources and capabilities that incumbents possess, they could be the accelerators of sustainability transitions if they recognised the necessity to reconfigure and transform [22,28,42,51]. Nevertheless, incumbents could serve as hybrid actors, linking niche innovations to incumbencies and thus altering and disrupting the regime [47], by taking part in the activities both in or between the regime and niche levels, bringing a new set of rules and requirements, even when operating in their own interest, with creation of links being their secondary concern [59], supporting the stable socio-technical system with constant fluctuation, never reaching a firm balance [2]. Hybrid organisations also possess the ability to gain profit not only with the quality of their services and products, but also to contribute to positive environmental and social change with their mission, with the profit not being the main objective, blurring the borders between for-profit and non-profit organisations [60], creating added value both to business and society [16]. This becomes achievable because hybrid organisations collaborating with the stakeholders of their social context strive to solve environmental and social issues through their practices and products, rather than just trying to reduce their negative impact [16,60]. Even when starting with incremental improvements, incumbents could create radical innovations, impacting the market when accelerated, and being adapted throughout the regime and niches [17]. Nonetheless, it is also necessary to keep in mind that these socio-technical systems are context-dependent and their individual transitions may vary depending on different circumstances [40].

Incumbents, being prominent actors in their sectors and regions, that can attribute their resources to influencing other members of the landscape developments, may take various roles in sustainable transitions, ranging from inhibitors to promoters and leaders:

- Resourceful opposers of sustainability transitions [4,42,56];
- Politically powerful environment shapers [40,42,44];
- Knowledge manipulators [38,45];
- Keepers of existing regimes [41,52];
- Incremental innovators [30,39,49–51];
- Promoters of new technologies and change [39,50,53];
- Hybrid actors [47,59].

Moreover, as some of the scientific literature on incumbent actors' role in the scientific literature on sustainability transitions mention the size and type of the firm (large manufacturers), these attributes are not necessarily important to address when looking at incumbent organisations. Incumbency of the organisation relies more on their strategical stance in the market, where they are established, prominent, with a set of network connections and stable positions that are not easily moved [61] by shifts during the development and perturbances of the landscape or the regime, allowing them to take any of the abovementioned roles.

2.2. Specificity of the Context of Transitioning Economy of a Post-Soviet Country

Lithuania is a country situated on the south-eastern side of the Baltic Sea with only 30 years of independence from the Soviet Union regime that has been shaping the country for 46 years. Lithuania is in a geopolitically active area with transit roads and most northern ice-free port of the Baltic Sea, which places the country in a position to successfully develop its economy. Currently, Lithuania belongs to the EU and NATO, providing both incentives and safety for successful independent development [25,62]. Sustainable development was started to be addressed in the early 2000s by being included into the National Strategy for Sustainable Development [63] and being recognised by companies creating Lithuanian Responsible Business Association (LAVA—Lietuvos Atsakingo Verslo Asociacija) in 2005, extending the work of the National Network of Responsible Business Enterprises (NAVĮT—Nacionalinis Atsakingo Verslo Įmonių Tinklas). Generally, the direction of companies represents the interests of consumers, shareholders, and other stakeholders, such as governments, policymakers, general society, etc. There are studies that investigate transitions of companies from planned to free market in the post-Soviet bloc [25,64], but not so much regarding sustainable development of the enterprises or countries. Search on the Web of Science Core Collection did not provide any results for keywords post-Soviet, sustainability transitions, organisation/business, and their synonyms. This might be due to the difference between sustainability and market transition, where sustainability is regarded as socio-technical, and market as a socio-economic system [25]. This notwithstanding, these systems are highly interlinked and co-dependent.

The Soviet Union did leave an imprint that might be felt in the contemporary society of the affected countries. The cultural legacy of this regime had multiple indirect negative impacts [23,65] in a broad variety of areas. Dawson [65] named several attributes linked with the mentality of post-Soviet countries: "passivity, circumspection, distrust, and a widespread indifference to environmental issues, pervading society, including governance systems, at multiple levels" (p. 56). Therefore, instead of gradual transition from one regime to another, these countries experienced shock-therapy [25,26] and some authors in the geography field refer to this transition more of as 'a form of transformation' [26,66,67]. However, experiencing these transitions, or transformations, possibly prepared the countries for sustainable transitions. Rodrigo et al. [23] conducted a study on transition dynamics and through comparison distinguished four groups of countries in terms of the way they approach sustainability issues: crossroaders, compliers, athletes, and laggards. The cluster of compliers contains fifteen ex-eastern bloc nations (except for Uzbekistan) that do particularly well in the quality of governance, but not in creating wealth cleanly, though they try to follow more sustainable paths, having not been exposed so much to the sustainable development concept. Growing economically strong, the compliers' cluster is not noted for effective and efficient energy management; however, they are addressing their effect on socio-environmental conditions, trying to improve it and comply with the standards raised by alliances such as the EU. Complier countries in the EU or those in the process of becoming a member must fulfil higher standards; however, they are not addressing their energy and CO_2 emission issues appropriately yet. Lithuania is presented as one of the exemplar cases complying with sustainability requirements through the use of some governmental pressures; it adopts improved industrial practices, takes moderate steps to reduce poverty and CO_2 emissions, but uses energy quite inefficiently, however, trying to develop more sustainably [23]. Citizens of the eastern post-Soviet bloc share the commonalities of inadequate energy use and a large proportion of the population is shown to struggle heating their houses in the cold periods and are experiencing energy poverty [68]; furthermore, are also undecided about the climate change [69].

Some challenges particular to Lithuania include limited availability of public transport; 10.6% of the population do not have access to indoor sanitation; the income of 20% of the richest people in the country was 7.1 times higher than 20% of the poorest people [70]; 20th place in the SDG Index of 2018 among 27 EU countries. Sustainable development principles in Lithuania are established through the main strategic planning documents of the country: Lithuania's Progress Strategy "Lithuania 2030"; 2014-2020 National Progress Programme; National Strategy for Sustainable Development adopted in 2003; the White Paper on Lithuanian Regional Policy prepared in 2017. The necessity for sustainable development is also mentioned in the Law on Territorial Planning of the Republic of Lithuania. However, it is noted that the topic of sustainable development lacks coherence and specificity in these documents, while the key issue in this regard is strategy: the National Strategy for Sustainable Development is more of a recommendatory nature [70]. Correspondingly, the Lithuanian National Sustainable Development Strategy has not been updated since 2011, whereas its implementation reports have not been submitted since 2014.

Nevertheless, membership in the EU had a major impact on Lithuania's development patterns. Post-Soviet countries that are members of the EU established market economy faster and performed wider-ranged reforms [62]. Integration in the EU can be seen as a stimulus for improving governance, having examples set by older members and absorbing sustainable development goals as part of the strategy [71]. Following the lead of more prominent EU countries that already have transitional experiences and competencies, should provide these transitioning economies with a boost for integrating sustainable development principles in their routines. The similarities between socio-technical and socio-economical system transformations should help transfer the multilevel concept of sustainable development to the market transition [25]. However, as the analysis above suggests, it is not to be expected from an incumbent regime to actively engage and invest in radical innovations as top-down initiatives do not intentionally generate niches and evolutionary bottom-up processes [18]. Therefore, it is important to address the initiative grounds of sustainability transitions, which could lay in incumbent actors in the context.

3. Materials and Methods

As a methodological approach, a single case study was selected to explore the incumbent actors' role in sustainability transitions in Lithuania. The majority of sustainability transitions' literature on incumbent actors is focused on energy sector. This landscape of Lithuania also has one of the main incumbent actors operating in energy sector. Therefore, information and examples of other incumbent actors will be drawn to better illustrate how companies in an emerging economy take the initiative and promote sustainable development. Five sustainability-oriented incumbent organisations in Lithuania were selected using nonprobability purposive sampling. First selection criterion was that the organisation had to be a part of the UN Global Compact initiative or a member of LAVA. The interviewed organisations could be categorised as incumbent actors in their sectors in Lithuania, being prominent and influential actors of their fields and active members of society, visible to the broader public, as a second criterion, where only two are large and have annual income above €100 million. After the second criterion was applied, an organisation that is not a part of UN Global Compact initiative or a member of LAVA emerged. Since it manages a national environmental brand label, it was included into the sample. Homogeneity of the sample is assured following the definition of incumbents that agrees with the research of Borghei [20], Sovacool et al. [28], Kungl [35], and Smink et al. [37], describing incumbents as established, prominent, and influential organisations. Characteristics of the interviewed organisations are provided in Table 2.

Table 2. Characteristics of respondents.

Organi-Sation	Sector	Size	Income, Mill € (2019)	Years Active	UN Global Compact	LAVA	Respondent
OrgC	Support services	Very small	0.3–0.5	18	+	-	Director
OrgD	Academic	Large	8–11	21	+	-	Sustainability coordinator
OrgE	Gas, water and multiutilities	Large	>100	12 (84)	+	+	Communication manager for sustainable development
OrgM	Support services	Small	1-3	17	-	-	Environment and sustainable development policy specialist
OrgR	Bank	Large	>100	27	+	+	Sustainability manager

Semi-structured interviews were conducted online using video conferencing means in Lithuanian during the period of November 2019 to August 2020. Interview questionnaire was built in consideration of the researches by Van der Byl and Slawinski [72] and McGrail et al. [73], revealing the organisations' stance towards sustainability. Representative questions are provided below:

- Could you tell when and on whose initiative social responsibility or sustainability became part of your organisation's strategy?
- Could you provide examples of social initiatives or solutions that your organisation is implementing?
- Could you provide examples of environmental initiatives or solutions that your organisation is implementing?
- Could you provide examples of economic initiatives or solutions that your organisation is implementing?
- What impact have the stakeholders had on the implementation of the social, environmental, and economic decisions you mentioned?
- What value and why did these social, environmental, and economic decisions bring to your organisation?

Transcripts of the interviews were made adopting a literal transcription strategy. Deductive qualitative coding [74] was applied to the transcripts where the codes (initiator; social, economic, environmental initiatives; stakeholders; value) were developed and complemented with the information from the interviews.

As additional information sources, corporate websites, and sustainability reports (of those organisations who report) where consulted using the same representative questions. This strategy was selected to complement the informants' knowledge with corporate sustainability communication, check for testimonies and validate the information.

4. Results

This section is dedicated to exploring the examples of five sustainability-oriented Lithuanian incumbents. Detailed stories and testimonies from the respondents are presented in Tables 3–7, providing information on why organisations took on the sustainability approach, their social, environmental, and economic initiatives, their stance towards stakeholders, and the value they see in being more sustainable.

Table 3. Testimony from the respondent on the sustainability of OrgD.

Whose initiative	• Environmental institute established in 1991 • Coordinated action since 2010 • Green higher education institution initiative started in 2012
Social initiatives	• No jeans month • Equal Opportunities Commission • Blood donation campaigns • Environment cleaning initiatives • First smart mobile app in Lithuania that helps the deaf to communicate with the hearing • Improving the infrastructure for all sorts of accessibility
Environmental initiatives	• Modernisation of heat and electricity facilities on the campus • Reduction in energy usage • Setting an example that we can use renewable energy resources • Development of technologies for disinfection of drinking water, protection against microbes • Development of an odour-sensitive sensor system that is able to detect air pollution
Stakeholders	• Organisation of events and open lectures for the community • Representatives participate in various TV and radio shows, provide press comments on various sustainability issues, spreading the message
Value	• Prestige, representation to the public • Raising awareness

Table 4. Testimony from the respondent on the sustainability of OrgM.

Whose initiative	• While consulting another company for a project, it became clear that we also have to become sustainable
Social initiatives	• Four-day week • Organising nature festivals • Family is brought to work • Staff and management renovated the office sustainably by themselves • Blood donation campaigns • Not engaging in sponsorship or philanthropy • Shaping the political level • Representing environmental interest of the public in court • Defending the public's right to nature
Environmental initiatives	• Renovating the office using green measures, such as adobe floor, or ecological wall paint • Calculating eco-footprint • Sorting, weighting their waste • Calculating trip costs using data from accounting, and deciding whether to take them, being aware where and what ecologic footprint they leave • Sustainable requirements for organised events, such as vegetarian food, eco-certified venues • Curating protected area brand • Encouraging all small producers in protected areas to put on a brand label • Training, teaching small producers marketing, sustainability, and responsibility
Economic initiatives	• Economic responsibility is understood more as product responsibility • Adapting to the economic capacity of partners • Economic responsibility is linked to the quality of the service • Quality assurance of services • Principle of fair price • Publicity
Stakeholders	• Raising the awareness of politicians • Raising the awareness of customers • Most influential interest group is the European Union and international organisations and their demands • The Green Deal affects all activities strongly • Other non-governmental environmental organisations are very important as partners • Influential academic institutions • Free meetings for business, where they get a lot information from us about environmental sustainability and we find out what is relevant in the market from them
Value	• Retention of valuable professionals • Calculating the footprint attracts clientele • Acting according to their words

Table 5. Testimony from the respondent on the sustainability of OrgC.

Whose initiative	• From the parent organisation • Global Compact pushed to document • Being open and transparent were the inner values when starting the organisation
Social initiatives	• Very protective of staff • Transparency • Public education on IT skills • Educating and training customers, businesses, and government agencies in the area of social responsibility • Talking about waste through creativity and art • Students and community interact with the waste management facility • Initiative of public emotional security project • Supplementary health insurance for employees
Environmental initiatives	• Educating public sector on green office solutions • Educating and organising excursions for schools regarding waste management • Collaborating with the university regarding the training of hazardous waste managers • Educating members and employees of governments and municipalities regarding waste management • Initiating emission transparency in waste management facilities
Economic initiatives	• Training economically active people and retirees to use the internet • Change in the remuneration of employees in public relations agencies
Stakeholders	• Being the initiator of sustainability initiatives • Avoid spreading disinformation
Value	• Employee loyalty and commitment of employees to adhere to and represent the principles and to adhere to the organisation's philosophy • Actions are in line with the philosophy, which is value for the clients

Table 6. Testimony from the respondent on the sustainability of OrgE.

Whose initiative	• Requirement of the Stock Exchange • The importance of social responsibility has risen from subsidiaries, making it a part of strategy of the group
Social initiatives	• Human rights, zero tolerance for discrimination • Domestic legislation • Code of ethics • Attract more women to technological specialties • Promoting a culture of safe work • Internal education campaigns regarding safety at work • Public education on energy • Employees tell about efficient energy use to the outside stakeholders • Maintaining personal contact with communities
Environmental initiatives	• Supporting and meeting environmental management standards • Promoting energy efficiency • Helping both the population and business to have a significant impact on the environment • Producing the majority of energy from renewable sources • Issuing Green Bonds that are used to finance green energy projects or energy efficiency projects • Smart energy club • Developing the network of electric car charging stations • Joining the United Nations Business Ambition for 1.5 °C
Economic initiatives	• Ethics, anti-corruption • Code of ethics • Line of trust • Annual tests for employees regarding anti-corruption • Testing partners' responsibility and sustainability • Work from home • Utility bills can be paid for free by customers using their self-service website • Customers experiencing financial difficulties can defer their payments • Positive impact on the Doing Business rating for Lithuania
Stakeholders	• Clear requirements from the state • The highest standards of transparency are required by the shareholders and investors • Management standards, benefits maximise employee retention • The media • Goal—to get stakeholders a little bit more involved so that they would provide feedback
Value	• Reputation improvement • Periodic public surveys to evaluate the reputation • No budget for publicising social responsibility

Table 7. Testimony from the respondent on the sustainability of OrgR.

Whose initiative	• Most strategic, political sustainability issues come from the parent company in Scandinavia • Very high level of enthusiasm and support for this topic comes from the CEO of the Lithuanian branch
Social initiatives	• Financial education • Promoting entrepreneurship • Building a sustainable society • Increasing awareness of sustainability • Employee volunteering initiative • Equal opportunities and diversity • Main sponsor of the human rights conference • Finance laboratory • Constant communication on personal finance, savings, pensions, investment, macroeconomics • A diverse programme for small business support • Endorsing and supporting women and entrepreneurs on innovations
Environmental initiatives	• Helping customers make more sustainable decisions • Green leasing • Special financing services for solar power • Investing in sustainable funds with high indices of sustainability • Sustainability risks are assessed in the same way as any other
Economic initiatives	• Start-up space for moving to sustainable innovation • Small Business Growing Programme • Promoting innovation in start-ups which then transforms into certain indicators of economic growth • Having sustainability criteria in the funding processes
Stakeholders	• Change of initiatives' partners is not common • Employees are always important • Shareholder and investor relations are always important • Institutional decision-makers, including the non-governmental sector, other companies with similar values with whom we can do some joint activities • Educational mission rather than responding to what clients would like • We proceed with external initiatives only after coordinating this with our internal audiences • Internal educational activities about sustainability
Value	• You will not be able to do successful business in any society that is doing badly • If a company is not sustainable, it will be harder to attract the younger generation of employees • The international investment community has already made it clear that all investments will be directed only at sustainable businesses • Failure to comply with legal requirements has the potential might lead to being expelled from the market • Public image

OrgD is a one of the top state-owned higher education institutions in Lithuania. This organisation presents its sustainability approach on its website; however, the report is outdated. The newest sustainability report can be found on the UN Global Compact website.

OrgM is a non-governmental non-profit organisation that operates in environmental protection area. OrgM reveals its mission on its website and on its sustainability report (as a part of its Activity report). Sustainability-oriented projects, activities, and initiatives can be found in both information sources (Table 4).

OrgC is a public relations organisation with affiliation to a larger company abroad. Double-checking the corporate website and sustainability report, little information could be found. Neither the website, nor the sustainability report mentioned stakeholder testimonies or their impact on the environment. A sustainability report is not on OrgC's website but it can be found on the UN Global Compact webpage (Table 5).

OrgE is a state-owned energy company, the most prominent incumbent in Lithuania's energy sector. The organisation has a very detailed approach to their sustainability strategy both on the website and in the sustainability report (which is a part of the annual financial report provided both on their website and that of the UN Global Compact) Table 6.

OrgR is one of the largest banks in Lithuania based on Scandinavian capital.

OrgR has sustainability as a part of their strategy, which is visible both on their website and sustainability report (which is provided alongside the Financial report). The report presents their activities, initiatives, and achieved results in great detail. The website shows consistency on all aspects of sustainability as well (Table 7).

5. Discussion

Interesting insights that illustrate Lithuania's regime and landscape levels could be drawn from the interviews and secondary data. A similarity shared by these organisations is that their top managers are involved in sustainability transitions and are active promoters of these ideas making them active members of transition towards sustainability both in their activity area and in the broader society. Some of these organisations became involved in sustainable movements around early 2000's participating in the development of the aforementioned association LAVA, making them prominent and long-lasting members of sustainability transitions in Lithuania. Nevertheless, respondents did notice that their clients, general society, and other stakeholders were still somewhat reluctant to address the sustainability issues, thus making it more complicated to initiate change and push these ideas beyond the company borders. Additionally, these organisations also face struggles from within. In example, OrgD encounters problems when broader and larger quantities of resources are required to implement the projects. In essence, looking at the interview data, website, and sustainability report, OrgD appears to have a fragmented approach towards sustainability; though, good intentions. OrgE has an abundance of statistical information on their report; however, some of the declared numbers that might be seen as controversial are not extensively elaborated upon, i.e., employee distribution by gender. OrgR's commitment to sustainable investment seems somewhat vague and unambitious, renouncing financing only to destructive activities, such as guns, coal energy, and those violating human rights.

Respondent from OrgM, who also is an expert in environmental and sustainability issues, noticed a huge difference in terms of corporate social responsibility: big organisations always have documents with clearly defined principles, and small organisations just have sustainability rising from within. Sustainability relates not so much to the organisation, but rather to the people that create the culture within. Those organisations that wanted to start a movement of change towards sustainability began looking for possibilities to collaborate, collect the positive examples from other companies, maybe even benchmark themselves against other sustainability-oriented enterprises, thus creating an informal network of socially responsible businesses in Lithuania (NAVĮT) in the beginning of 2000″s. However, OrgC's respondent admits that when looking globally, in addressing sustainability, Lithuania still lags four or five years behind. Therefore, it is necessary to be proactive in discovering sustainability and managing risks in the organisations and other social constructs, to have a steppingstone to move forward and communicate and consult, and include all stakeholders, according to the respondent of OrgM. Though the research may not yet show a strong willingness to pay extra for a sustainability focus in a product or service, that trend is growing, as noted by the respondent of OrgR. Organisations are also more prone not to push sustainable ideas actively to the clients, but to focus on philanthropy and sponsorship as an act of sustainability (respondents of OrgM and OrgC). There are organisations using social events (like marathons) to gather sensitive data on people participating in them and then promoting their products directly (OrgM's respondent). Nonetheless, there are companies that are unequivocally profit oriented and do not have, want to have, or discuss sustainability. The majority of those who address sustainability still see this as a responsibility of the communication department. However, respondent of OrgR notes that unsustainable businesses will not have a place in the future and today, sustainability is becoming a business topic, an important topic. The enthusiasm that is being expressed by the respondents of the analysed organisations could be illustrated by the reflection of OrgM's respondent: "sustainability and social responsibility are not the implementation of some standard, they are not the implementation of some postulates, they are not a policy, they are not a principle. It's that jazz that should drive that organisation and should just be fun, becoming a part of that organisation." What seems to be clear from the literature and empirical analysis, is that collaboration between organisations is necessary to tackle the issues of sustainability [16], with key success factors for adapting sustainability at the core of the corporation being structure, culture, leadership, communication, employee qualifications and motivation, and management control [48,75].

Moreover, a trend could be drawn from this small sample of organisations that prominent organisations can dedicate a broad variety of resources to address their corporate sustainability and the message they send, which is visible in their websites and sustainability reports. Naturally, this does not imply that large incumbents are more interested in what message to send than actually living up to the promise. Larger organisations can allocate more resources to various activities, where smaller organisations invest their limited resources to the actual activities. Additionally, large incumbents ascribe greater value to SDGs in their communication sources, structuring their messages accordingly: which SDGs are important to them, to which SDGs do they contribute the most.

This overview on Lithuania's incumbent actors provides insights that support the emerging ideas that incumbents are not only stalling the sustainability transition and anticipating change—they can be active promoters of sustainability ideas who actively motivate their stakeholders to take the sustainability path and try to educate the larger society, taking on initiatives and leading the change, complementing the research of Berggren et al. [50], Hörisch [39], and Markard [53]. Additionally, OrgM reveals traits that are inherent to hybrid actors, linking niche innovations of small enterprises with other market players [47] and maintaining economic profit while adding value to society and the environment [16,60]. OrgE provides an example of positive change in the policy of energy sector acting as a politically powerful environment shaper, bringing up the transition from inside and supporting decentralisation and green energy ideas, taking action in building a more sustainable energy supply, whilst other scientific researches provide a more negative depiction of state-owned enterprises [40,64]. Moreover, even though incumbents as incremental innovators are often referred via technological solutions, OrgR poses as an example of incremental developer of society, taking small steps for the education of communities in various directions.

However, the major portion of previous literature on incumbents in sustainability transitions relies on energy sector examples, whereas the sustainability transitions in the post-Soviet context focus on urban/rural development and/or agriculture. Consequently, comparison with other sectors is complicated. Insights on business development and impact on tackling sustainability issues remain an underexplored area both in the case of sustainability transitions and post-Soviet country context. Interestingly, Chatzimentor et al. [24] noted that studies from eastern and post Soviet countries were often left at the periphery of peer-reviewed journals, having lower-acceptance rates [76,77]. Nonetheless, this research, like the larger part of other transition studies, is designed to be more illustrative and exploratory than a systematic research, possessing elements of interpretation and creativity in its methodology [51]. This paper illustrates how organisations of a transitioning economy take on the initiatives to become more sustainable, empowering their stakeholders to take a part in the transition towards sustainability, when the government still lags behind, trying to meet the requirements set by foreign partners and alliances.

6. Conclusions

The aim of the paper was to explore what roles do the incumbent actors take in the sustainability transitions. The scientific literature review suggested seven roles: resourceful opposers of sustainability transitions; politically powerful environment shapers; knowledge manipulators; keepers of existing regimes; incremental innovators; promoters of new technologies and change; hybrid actors. The analysed sample of five sustainability-oriented incumbent organisations in a transitioning economy revealed that they all are promoters of change; additionally, one could act as a hybrid actor; two organisations highlighted positive attributes of two roles, one being politically powerful shaper and the other—an incremental innovator. Moreover, the incumbent actors that take on the role of promoting sustainability are emerging in Lithuania and their positive examples supplement the shifting views on incumbents in the scientific literature. This paper, as an overview of the situation in an emerging economy, provides ideas and a steppingstone for future research regarding the role of incumbents that could be used to advance and facilitate the management of sustainability transitions.

Limitations of this paper include a purely qualitative approach when investigating examples of incumbent actors, drawing only from data provided by the organisations: interviews with respondents who themselves were very interested and vested in sustainability ideas and movement, corporate sustainability reports, and websites. One of the ways to further develop this research would be to explore the organisations more thoroughly, looking within these organisations for clues whether there was interest from certain individuals, or all members of organisations supported the sustainability ideas. MLP is better suited for describing events than looking at causal relationships [14,18]; therefore, this research, aimed at exploring the role of incumbent actors, is limited at the level of incumbent actors' initiatives, with only slight insights into why they transition towards sustainability and their certain manner, contributing more to a geographical perspective of the sustainable transitions. It could also be interesting to examine whether transition from planned to market economy provided any advantage transitioning towards sustainability. Correspondingly, an investigation of how the post-Soviet country mentality shaped the actions of incumbents could provide fruitful insights into the pathways these organisations are taking and why. Due to early stages of sustainable development idea implementation in strategies in transitioning economies, this might be impossible to assess yet.

Funding: This research received no external funding.

Institutional Review Board Statement: Not applicable.

Informed Consent Statement: Informed consent was obtained from all interviewees involved in the study.

Data Availability Statement: The data presented in this study are available on request from the corresponding author. The data are not publicly available due to privacy restrictions and anonymity of the respondents.

Acknowledgments: I would like to express my gratitude to the communities of the 27th Annual Conference of the International Sustainable Development Research Society (ISDRS) and the 6th NEST conference and the Network for Early career researchers in Sustainability Transitions for the possibility to present my research and receive feedback that allowed me to develop this paper. I would also like to thank the anonymous reviewers who have brought wonderful insights to improve my research.

Conflicts of Interest: The author declares no conflict of interest.

References

1. Ramanauskaitė, J. Incumbents in sustainability transitions in the context of transitioning economy: An onlook of incumbent actors' initiatives. In Proceedings of the ISDRS 2021: The 27th International Sustainable Development Research Society Conference: Accelerating the Progress towards the 2030 SDGs in Times of Crisis, Östersund, Sweden, 13–15 July 2021; Johansson, C., Mauerhofer, V., Eds.; Mittuniversitetet: Östersund, Sweden, 2021; p. 392.
2. van Mossel, A.; van Rijnsoever, F.J.; Hekkert, M.P. Navigators through the storm: A review of organization theories and the behavior of incumbent firms during transitions. *Environ. Innov. Soc. Transit.* **2018**, *26*, 44–63. [CrossRef]
3. Strøm-Andersen, N. Incumbents in the Transition Towards the Bioeconomy: The Role of Dynamic Capabilities and Innovation Strategies. *Sustainability* **2019**, *11*, 5044. [CrossRef]
4. Hansen, T.; Coenen, L. Unpacking resource mobilisation by incumbents for biorefineries: The role of micro-level factors for technological innovation system weaknesses. *Technol. Anal. Strateg. Manag.* **2017**, *29*, 500–513. [CrossRef]
5. Birch, K.; Levidow, L.; Papaioannou, T. *Sustainable Capital?* The Neoliberalization of Nature and Knowledge in the European "Knowledge-based Bio-economy." *Sustainability* **2010**, *2*, 2898–2918. [CrossRef]

6. Markard, J.; Raven, R.; Truffer, B. Sustainability transitions: An emerging field of research and its prospects. *Res. Policy* **2012**, *41*, 955–967. [CrossRef]
7. Schlaile, M.P.; Urmetzer, S. Transitions to Sustainable Development. In *Decent Work and Economic Growth*; Leal Filho, W., Azul, A.M., Brandli, L., Özuyar, P.G., Wall, T., Eds.; Springer International Publishing: Cham, Switzerland, 2019; pp. 1–16. ISBN 978-3-319-71058-7.
8. Lyytimäki, J.; Vikström, S.; Furman, E. Voluntary participation for sustainability transition: Experiences from the 'Commitment to Sustainable Development 2050'. *Int. J. Sustain. Dev. World Ecol.* **2019**, *26*, 25–36. [CrossRef]
9. Fuenfschilling, L.; Truffer, B. The structuration of socio-technical regimes—Conceptual foundations from institutional theory. *Res. Policy* **2014**, *43*, 772–791. [CrossRef]
10. Geels, F.W. Ontologies, socio-technical transitions (to sustainability), and the multi-level perspective. *Res. Policy* **2010**, *39*, 495–510. [CrossRef]
11. Geels, F.W. Technological transitions as evolutionary reconfiguration processes: A multi-level perspective and a case-study. *Res. Policy* **2002**, *31*, 1257–1274. [CrossRef]
12. Smith, A.; Voß, J.-P.; Grin, J. Innovation studies and sustainability transitions: The allure of the multi-level perspective and its challenges. *Res. Policy* **2010**, *39*, 435–448. [CrossRef]
13. Hannon, M. Co-Evolution of Innovative Business Models and Sustainability Transitions: The Case of the Energy Service Company (ESCo) Model and the UK Energy System. Ph.D. Thesis, University of Leeds, Leeds, UK, 2012.
14. Svensson, O.; Nikoleris, A. Structure reconsidered: Towards new foundations of explanatory transitions theory. *Res. Policy* **2018**, *47*, 462–473. [CrossRef]
15. Frantzeskaki, N.; Loorbach, D.; Meadowcroft, J. Governing societal transitions to sustainability. *Int. J. Sustain. Dev.* **2012**, *15*, 19. [CrossRef]
16. Loorbach, D.; Wijsman, K. Business transition management: Exploring a new role for business in sustainability transitions. *J. Clean. Prod.* **2013**, *45*, 20–28. [CrossRef]
17. Schaltegger, S.; Lüdeke-Freund, F.; Hansen, E.G. Business Models for Sustainability. *Organ. Environ.* **2016**, *29*, 264–289. [CrossRef]
18. Geels, F.W.; Schot, J. Typology of sociotechnical transition pathways. *Res. Policy* **2007**, *36*, 399–417. [CrossRef]
19. Apajalahti, E.-L. *Large Energy Companies in Transition—From Gatekeepers to Bridge Builders*; Aalto University: Helsinki, Finland, 2018.
20. Borghei, B.B. *Incumbent Actors in Sectoral Transformations Towards Sustainability: A Sociotechnical Study of the European Heavy Commercial Vehicles Sector*; Linköping University Electronic Press: Linköping, Sweden, 2018.
21. Dewald, U.; Achternbosch, M. Why did more sustainable cements failed so far? Disruptive innovations and their barriers in a basic industry. *Environ. Innov. Soc. Transit.* **2016**, *19*, 15–30. [CrossRef]
22. Hockerts, K.; Wüstenhagen, R. Greening Goliaths versus emerging Davids—Theorizing about the role of incumbents and new entrants in sustainable entrepreneurship. *J. Bus. Ventur.* **2010**, *25*, 481–492. [CrossRef]
23. Rodrigo, P.; Muñoz, P.; Wright, A. Transitions dynamics in context: Key factors and alternative paths in the sustainable development of nations. *J. Clean. Prod.* **2015**, *94*, 221–234. [CrossRef]
24. Chatzimentor, A.; Apostolopoulou, E.; Mazaris, A.D. A review of green infrastructure research in Europe: Challenges and opportunities. *Landsc. Urban Plan.* **2020**, *198*, 103775. [CrossRef]
25. Fischer, D. Comparing transitions: Insights from the economic transition processes in former socialist countries for sustainability transitions. *Osteuropa-Wirtschaft* **2010**, *55*, 289–310.
26. Brown, G.; Kraftl, P.; Pickerill, J.; Upton, C. Holding the Future Together: Towards a Theorisation of the Spaces and Times of Transition. *Environ. Plan. A Econ. Sp.* **2012**, *44*, 1607–1623. [CrossRef]
27. Holtz, G.; Brugnach, M.; Pahl-Wostl, C. Specifying "regime"—A framework for defining and describing regimes in transition research. *Technol. Forecast. Soc. Chang.* **2008**, *75*, 623–643. [CrossRef]
28. Sovacool, B.K.; Turnheim, B.; Martiskainen, M.; Brown, D.; Kivimaa, P. Guides or gatekeepers? Incumbent-oriented transition intermediaries in a low-carbon era. *Energy Res. Soc. Sci.* **2020**, *66*, 101490. [CrossRef]
29. Johnstone, P.; Stirling, A.; Sovacool, B. Policy mixes for incumbency: Exploring the destructive recreation of renewable energy, shale gas 'fracking,' and nuclear power in the United Kingdom. *Energy Res. Soc. Sci.* **2017**, *33*, 147–162. [CrossRef]
30. Hengelaar, G. *The Proactive Incumbent: Holy Grail or Hidden Gem?* Erasmus University Rotterdam: Rotterdam, The Netherlands, 2017.
31. Chandy, R.K.; Tellis, G.J. The Incumbent's Curse? Incumbency, Size, and Radical Product Innovation. *J. Mark.* **2000**, *64*, 1–17. [CrossRef]
32. Henderson, R. Underinvestment and Incompetence as Responses to Radical Innovation: Evidence from the Photolithographic Alignment Equipment Industry. *RAND J. Econ.* **1993**, *24*, 248. [CrossRef]
33. Mitchell, W. Dual clocks: Entry order influences on incumbent and newcomer market share and survival when specialized assets retain their value. *Strateg. Manag. J.* **1991**, *12*, 85–100. [CrossRef]
34. Mitchell, W.; Singh, K. Death of the Lethargic: Effects of Expansion into New Technical Subfields on Performance in a Firm's Base Business. *Organ. Sci.* **1993**, *4*, 152–180. [CrossRef]
35. Kungl, G. Stewards or sticklers for change? Incumbent energy providers and the politics of the German energy transition. *Energy Res. Soc. Sci.* **2015**, *8*, 13–23. [CrossRef]

36. Fligstein, N.; McAdam, D. *A Theory of Fields*; Oxford University Press: New York, NY, USA, 2012; ISBN 0199859957.
37. Smink, M.M.; Hekkert, M.P.; Negro, S.O. Keeping sustainable innovation on a leash? Exploring incumbents' institutional strategies. *Bus. Strateg. Environ.* **2015**, *24*, 86–101. [CrossRef]
38. Turnheim, B.; Sovacool, B.K. Forever stuck in old ways? Pluralising incumbencies in sustainability transitions. *Environ. Innov. Soc. Transit.* **2020**, *35*, 180–184. [CrossRef]
39. Hörisch, J. How business actors can contribute to sustainability transitions: A case study on the ongoing animal welfare transition in the German egg industry. *J. Clean. Prod.* **2018**, *201*, 1155–1165. [CrossRef]
40. Lindberg, M.B.; Markard, J.; Andersen, A.D. Policies, actors and sustainability transition pathways: A study of the EU's energy policy mix. *Res. Policy* **2019**, *48*, 103668. [CrossRef]
41. Matschoss, K.; Heiskanen, E. Innovation intermediary challenging the energy incumbent: Enactment of local socio-technical transition pathways by destabilisation of regime rules. *Technol. Anal. Strateg. Manag.* **2018**, *30*, 1455–1469. [CrossRef]
42. Geels, F.W. Socio-technical transitions to sustainability: A review of criticisms and elaborations of the Multi-Level Perspective. *Curr. Opin. Environ. Sustain.* **2019**, *39*, 187–201. [CrossRef]
43. Geels, F.W. Regime Resistance against Low-Carbon Transitions: Introducing Politics and Power into the Multi-Level Perspective. *Theory Cult. Soc.* **2014**, *31*, 21–40. [CrossRef]
44. Walz, R.; Köhler, J. Using lead market factors to assess the potential for a sustainability transition. *Environ. Innov. Soc. Transit.* **2014**, *10*, 20–41. [CrossRef]
45. Sovacool, B.K.; Brisbois, M.-C. Elite power in low-carbon transitions: A critical and interdisciplinary review. *Energy Res. Soc. Sci.* **2019**, *57*, 10. [CrossRef]
46. Hess, D.J. Sustainability transitions: A political coalition perspective. *Res. Policy* **2014**, *43*, 278–283. [CrossRef]
47. Kallio, L.; Heiskanen, E.; Apajalahti, E.-L.; Matschoss, K. Farm power: How a new business model impacts the energy transition in Finland. *Energy Res. Soc. Sci.* **2020**, *65*, 101484. [CrossRef]
48. Wolff, S.; Brönner, M.; Held, M.; Lienkamp, M. Transforming automotive companies into sustainability leaders: A concept for managing current challenges. *J. Clean. Prod.* **2020**, *276*, 124179. [CrossRef]
49. Schumpeter, J.A. *Capitalism, Socialism, and Democracy*, 3rd ed.; Harper & Brothers: New York, NY, USA, 1950.
50. Berggren, C.; Magnusson, T.; Sushandoyo, D. Transition pathways revisited: Established firms as multi-level actors in the heavy vehicle industry. *Res. Policy* **2015**, *44*, 1017–1028. [CrossRef]
51. Geels, F.W. The multi-level perspective on sustainability transitions: Responses to seven criticisms. *Environ. Innov. Soc. Transit.* **2011**, *1*, 24–40. [CrossRef]
52. Bosman, R.; Loorbach, D.; Frantzeskaki, N.; Pistorius, T. Discursive regime dynamics in the Dutch energy transition. *Environ. Innov. Soc. Transit.* **2014**, *13*, 45–59. [CrossRef]
53. Markard, J. The life cycle of technological innovation systems. *Technol. Forecast. Soc. Chang.* **2020**, *153*, 119407. [CrossRef]
54. Mossberg, J.; Söderholm, P.; Hellsmark, H.; Nordqvist, S. Crossing the biorefinery valley of death? Actor roles and networks in overcoming barriers to a sustainability transition. *Environ. Innov. Soc. Transit.* **2018**, *27*, 83–101. [CrossRef]
55. Geels, F.W. Processes and patterns in transitions and system innovations: Refining the co-evolutionary multi-level perspective. *Technol. Forecast. Soc. Chang.* **2005**, *72*, 681–696. [CrossRef]
56. Kuokkanen, A.; Nurmi, A.; Mikkilä, M.; Kuisma, M.; Kahiluoto, H.; Linnanen, L. Agency in regime destabilization through the selection environment: The Finnish food system's sustainability transition. *Res. Policy* **2018**, *47*, 1513–1522. [CrossRef]
57. Kungl, G.; Geels, F.W. Sequence and alignment of external pressures in industry destabilisation: Understanding the downfall of incumbent utilities in the German energy transition (1998–2015). *Environ. Innov. Soc. Transit.* **2018**, *26*, 78–100. [CrossRef]
58. Mignon, I.; Kanda, W. A typology of intermediary organizations and their impact on sustainability transition policies. *Environ. Innov. Soc. Transit.* **2018**, *29*, 100–113. [CrossRef]
59. Elzen, B.; van Mierlo, B.; Leeuwis, C. Anchoring of innovations: Assessing Dutch efforts to harvest energy from glasshouses. *Environ. Innov. Soc. Transit.* **2012**, *5*, 1–18. [CrossRef]
60. Haigh, N.; Hoffman, A.J. Hybrid Organizations: The Next Chapter in Sustainable Business. *SSRN Electron. J.* **2011**, *41*. [CrossRef]
61. Black, J.; Hashimzade, N.; Myles, G. *A Dictionary of Economics. Incumbent Firm*, 3rd ed.; Oxford University Press: New York, NY, USA, 2009; ISBN 9780199237043.
62. Cameron, D.R. Creating Market Economies after Communism: The Impact of the European Union. *Post-Sov. Aff.* **2009**, *25*, 1–38. [CrossRef]
63. Lietuvos Respublikos Vyriausybė Nutarimas Nr. 1160. Dėl Nacionalinės Darnaus Vystymosi Strategijos Patvirtinimo ir Igyvendinimo. *Valstyb. Žinios.* 2003. Available online: https://e-seimas.lrs.lt/portal/legalAct/lt/TAD/TAIS.217644/WPqyZkDuCy (accessed on 25 October 2021).
64. Tõnurist, P. Framework for analysing the role of state owned enterprises in innovation policy management: The case of energy technologies and Eesti Energia. *Technovation* **2015**, *38*, 1–14. [CrossRef]
65. Dawson, L. *Unravelling Sustainability: The Complex Dynamics of Emergent Environmental Governance and Management Systems at Multiple Scales*; Stockholm University: Stockholm, Sweden, 2019.
66. Lynn, N.J. Geography and Transition: Reconceptualizing Systemic Change in the Former Soviet Union. *Slavic Rev.* **1999**, *58*, 824–840. [CrossRef]

67. Smith, A. From Convergence to Fragmentation: Uneven Regional Development, Industrial Restructuring, and the 'Transition to Capitalism' in Slovakia. *Environ. Plan. A Econ. Sp.* **1996**, *28*, 135–156. [CrossRef]
68. Bouzarovski, S.; Tirado Herrero, S.; Petrova, S.; Frankowski, J.; Matoušek, R.; Maltby, T. Multiple transformations: Theorizing energy vulnerability as a socio-spatial phenomenon. *Geogr. Ann. Ser. B Hum. Geogr.* **2017**, *99*, 20–41. [CrossRef]
69. Ferenčuhová, S. Not so global climate change? Representations of post-socialist cities in the academic writings on climate change and urban areas. *Eurasian Geogr. Econ.* **2020**, *61*, 686–710. [CrossRef]
70. Punytė, I.; Simonaitytė, K. *Darnaus Vystymosi Tikslų Integravimas į Kompleksinius Teritorinius Planus. Darnaus Vystymosi Tikslai ir Planavimo Sistema Lietuvoje: Esamos Situacijos Analizė*; Kurk Lietuvai: Vilnius, Lithuania, 2018.
71. Leal Filho, W.; Platje, J.; Gerstlberger, W.; Ciegis, R.; Kääriä, J.; Klavins, M.; Kliucininkas, L. The role of governance in realising the transition towards sustainable societies. *J. Clean. Prod.* **2016**, *113*, 755–766. [CrossRef]
72. Van der Byl, C.A.; Slawinski, N. Embracing Tensions in Corporate Sustainability: A Review of Research From Win-Wins and Trade-Offs to Paradoxes and Beyond. *Organ. Environ.* **2015**, *28*, 54–79. [CrossRef]
73. McGrail, S.; Halamish, E.; Teh-White, K.; Clark, M. Diagnosing and anticipating social issue maturation: Introducing a new diagnostic framework. *Futures* **2013**, *46*, 50–61. [CrossRef]
74. Žydžiūnaitė, V.; Sabaliauskas, S. *Kokybiniai Tyrimai: Principai ir Metodai: Vadovėlis Socialinių Mokslų Studijų Programų Studentams*; Vaga: Vilnius, Lithuania, 2017.
75. Engert, S.; Baumgartner, R.J. Corporate sustainability strategy—bridging the gap between formulation and implementation. *J. Clean. Prod.* **2016**, *113*, 822–834. [CrossRef]
76. Rotter, A.; Gostincar, C. A Defense of Eastern European Science. *Science* **2014**, *343*, 839. [CrossRef]
77. Sutcliffe, L.M.E.; Batáry, P.; Kormann, U.; Báldi, A.; Dicks, L.V.; Herzon, I.; Kleijn, D.; Tryjanowski, P.; Apostolova, I.; Arlettaz, R.; et al. Harnessing the biodiversity value of Central and Eastern European farmland. *Divers. Distrib.* **2015**, *21*, 722–730. [CrossRef]

MDPI

Article

Decisions by Key Office Building Stakeholders to Build or Retrofit Green in Toronto's Urban Core

Prescott C. Ensign [1,*], Shawn Roy [2] and Tom Brzustowski †

1 Lazaridis School of Business and Economics, Wilfrid Laurier University, Waterloo, ON N2L3C7, Canada
2 Offsets and Partner Engagement, MDA, Ottawa, ON K2E8B2, Canada; sroy046@gmail.com
* Correspondence: ensign@wlu.ca
† Deceased 19 June 2020. Officer of the Order of Canada, Fellow of the Canadian Academy of Engineering, Fellow of the Royal Society of Canada, First provost of the University of Waterloo, former deputy minister in the Government of Ontario, president of the Natural Sciences and Engineering Research Council for over 10 years, senior advisor to the Institute for Quantum Computing. Honorary member of the Royal Military College of Canada as well as honorary degrees from University of Guelph, Ryerson Polytechnic University, University of Waterloo, McMaster University, University of Alberta, Université d'Ottawa, École Polytechnic de Montréal, Concordia University, Brock University, University of Victoria, York University, University of Northern British Columbia, Carleton University and Lakehead University.

Abstract: The environmental impact of greenhouse gas emissions from buildings—especially in global cities such as Toronto—is well documented. Green mitigation of new and existing buildings has also been researched. Few studies, however, have focused on the decision to build or retrofit green. Are key stakeholders in Toronto's office building sector aligning their decisions to achieve sustainable environmental goals? Do they support LEED certification regardless of the impact on market valuation? Are tenants willing to pay higher rents in LEED office buildings? The study first obtained data on 16 LEED and 52 conventional buildings to determine if LEED certification has a significant impact on net asking rent. Pearson correlation and linear regression analysis did not find LEED certification to be statistically significant in explaining the variance in net asking rent (market value). The second stage included interviews with senior executives engaged in Toronto's office building sector. The expert informtabants were asked to assess if financial drivers are the deciding factor in decisions to pursue LEED certification. They concurred that LEED certification is not the primary driver. It is a combination of numerous factors that overall have an impact on a firm's financial bottom line.

Keywords: green buildings; LEED certification; real estate development process; drivers of sustainability

Citation: Ensign, P.C.; Roy, S.; Brzustowski, T. Decisions by Key Office Building Stakeholders to Build or Retrofit Green in Toronto's Urban Core. *Sustainability* **2021**, *13*, 6969. https://doi.org/10.3390/su13126969

Academic Editor: Ioannis Nikolaou

Received: 3 March 2021
Accepted: 31 May 2021
Published: 21 June 2021

1. Introduction

Global warming and climate change are well-established facts that call for mitigation and adaptation actions in the building sector [1]. Some argue that the potential to address climate change by green technology is underestimated. Action is needed both in new construction and deep energy retrofits in existing buildings [2–5]. It is especially important in Canada where the building sector accounts for 12 percent of overall greenhouse gas (GHG) emissions [6]. Toronto's GHG emissions emitted in 2018 from residential, commercial and industrial buildings totaled 55 percent or 8.9 megatonnes, an increase of 13 percent compared to 2017 [7]. Toronto's City Council declared a climate emergency on October 2, 2019, joining a global call to recognize the urgency of the climate crisis. To underscore this, the City Council adopted a stronger emissions reduction target for Toronto—net zero by 2050 or sooner [7].

The federal and provincial governments in Canada have also acted on climate change by setting carbon mitigation targets for the year 2030 based on the Paris Climate Agreement and the Pan-Canadian Framework. In December 2020 Ottawa announced plans to raise the federal carbon tax to Canadian dollar (CAD) 170 a tonne by 2030, up from the current

CAD 30 a tonne [8]. An increase in the carbon tax acts as a negative reinforcement to encourage energy efficiency improvements; this has proved to be successful in mitigating GHG emission [2]. The objective of these efforts is to link the efficiency and productivity of buildings to the goal of achieving the 2030 United Nations Sustainable Development Goals (SDGs) [9]. This will not be easy to achieve and leadership matters.

Interest in sustainable and ecologically benign green office buildings has gained momentum over the past 20 years in Canada. One indicator is the growth in LEED—Leadership in Energy and Environmental Design—certified projects in Canada. LEED is one of the most widely recognized green building rating systems in the world. Prior to 2005 Canada had less than 200 buildings registered with LEED [10]. According to the Canada Green Building Council (CaGBC) by the end of 2018 Canada had a cumulative total of 3254 LEED certified projects with a total of 46.81 million gross square meters [11]. The significant increase in projects indicates a strong commitment to buildings that promote a healthier, more sustainable future. In order for Canada to continue its position as a leader in the adoption of green commercial construction projects it is important to understand the factors that contribute to this growth in sustainable building development. Toronto—Canada's largest office market—is an ideal test site for understanding this issue.

The purpose of this study is to address the question: Why are key office building stakeholders—real estate developers, institutional investors and owner landlords—deciding to build or retrofit green in Toronto's urban core? The particular interest for this study is to determine (1) the impact of LEED certification on office building market rent valuation, and (2) to identify the financial and non-financial factors that influence the initial decision to build or retrofit green.

2. Conceptual Framework for the Study

Green construction and retrofitting of buildings will play an important role in reversing climate change. This calls for mitigation and adaptation actions at the local, national, regional and global level. This exploratory study looks at the decisions by key stakeholders to build or retrofit office buildings green in Toronto's prime commercial office market. The framework for this study is developed from an examination of contextual studies and process theories drawn from a number of different disciplines that address this decision.

2.1. Place-Based Context

Contextual place-based studies on the construction and retrofitting of commercial office buildings adds to our global understanding for green advocacy. More importantly, location specific research can lead to developing and implementing a range of approaches for green intervention that result in sustainable development strategies. This study helps to fill the gap in research on cities such as Toronto (central urban population 2.9 m/total greater Toronto area population 6.8 m) that have successfully transitioned to a post-industrialised economy. These cities have a large population and central business district (CBD) urban core with a prime commercial office building market.

Based on a review of studies that focus specifically on green office buildings in large urban settings, there are three studies that provide insight. These studies looked at buildings in Singapore, Milan and Hong Kong. The 2009 study of 400 users of commercial buildings in the city-state of Singapore (5.7 m) found that office occupants were reluctant to invest in and/or occupy a green building (GB) [12]. One wonders how these respondents would respond twelve years later. The study of Mangialardo et al. [13] published in 2018 analysed 55 office building projects in Milan (3.1 m), Italy's most flourishing real estate market. Their findings indicate that rent premiums and higher prices are generated in LEED certified properties. Additionally, green properties are absorbed by the market in less than half the time when compared to those without green certification [14]. The 2019 study that looked at the city of Hong Kong (7.5 m) by Wadu Mesthrige and Chan [15] found that developer and investor stakeholders were uncertain and sceptical about the financial rewards of green certification. The quantitative study tested the factual basis for these inhibiters with the actions of the real estate market. A hedonic-model of rent (market) and building attributes including green certification was used to assess their sample of 67 green and non-green prime office buildings. The findings suggest that: green certified office space added value to the property; and tenants were willing to pay 10.9% higher rent than for comparable non-green space. These studies help to support the view that green real estate development can and does make a significant contribution to the economic, environmental and social sustainability of global cities such as Milan, Hong Kong and Toronto.

On the issue of place/location, the Simons et al. [16] study on the US city of Cleveland (1.8 m) points to the reality that land in a city's urban core is finite and most of it already has existing buildings. When a footprint in the urban core comes on the market a number of developers with competing uses vie for the property. Each developer's plan will have varying degrees of positive and negative environmental, social and economic impact. The point is that a green building project involves numerous stakeholders and potentially incompatible concerns.

2.2. Decision Process to Build Green

Our review of relevant literature included studies that examined the decision process in a green commercial office building project. Graaskamp [17] raises the question: 'Who participates in the decision process?' The answer is very few at the initial decision point, compared to the large number of persons who are involved in designing, selecting systems, engineering and government approval. This issue is investigated by Goubran and Cucuzzella [18]. Their Province of Québec study (2019) looked at how localised projects could integrate the 2030 SDGs Agenda into the building design team process. This team activity is moved downstream from the upstream process where the critical decision is made to build green. This initiating decision is an outcome of a logical process [19].

One widely used theoretical process approach for examining decision making at the micro-level is the entrepreneurial process model suggested by Timmons et al. [20]. It provides a framework for understanding the dynamics of aligning the three critical components—opportunity, resources and team—that will drive the decision. This approach would view the key stakeholders (real estate developers) as acting such as serial entrepreneurs who approach each new project as though it is the formation of a new venture.

Another widely used process framework for understanding decision-making is to look at a firm's business model as dynamic that is adjusting, balancing and evolving over time. As Teece notes:

> 'The essence of a business model is in defining the manner by which the enterprise delivers value to customers, entices customers to pay for value, and converts those payments to profit. It thus reflects management's hypothesis about what customers want, how they want it, and how the enterprise can organize to best meet those needs, get paid for doing so, and make a profit' [21].

A firm's business model is reflected in the key internal stakeholders' decisions and motivations related to green office buildings [22]. Traditional business models view value creation in a one direction flow as opposed to business models that focus on sustainability [23]. A change to a firm's business model is not an insignificant change in the assumptions that guide the organization in shaping its value proposition to clients, balancing resources in its ecosystem and generates cash flow [20]. The business model canvas [24] is especially useful for analysis of the decisions that are made at the firm level, but it requires access to considerable internal information. Joyce and Paquin's [25] triple layered (economic, environmental and social) business model canvas study is a particularly useful approach.

However, Comin et al. [26] note that access to sufficient information is needed. Few sustainable business models presented in the literature offer explicit ways of operationalizing the proposed models. Preghenella and Battistella's [23] bibliographic review presents an overarching theoretical framework for mapping the research streams of business models for sustainability. Based on their findings they concluded that the sustainability role in a company business model is still undefined in the research. This points to the benefit of an exploratory study that focuses on the decision process.

Perhaps the best approach is to look at the field of real estate that has its own practitioner devised and applied process model. It takes into consideration that: (1) the metrics of each urban core real estate project are unique; and (2) the development process is often more political than economic. In a free market economy, each new building project is an opportunity for society and key stakeholders to negotiate, debate and reconsider who pays, who benefits, who risks and who has standing to participate in the decision process [17]. Whether conventional or green, most real estate development processes are similar. Zaccack [27] presents it conceptually as a six-stage process: idea inception, feasibility, pre-construction, construction, operation and measure. The first three stages of the process are directly applicable to this study. During the idea inception stage, the developer will do initial studies on the market, design, zoning, and financing to understand the possibilities available. This stage is typically conducted internally (developer, owner and investor) and rarely includes outside consultants. During the feasibility stage, external consultants are hired beginning with an architect and other professionals including LEED. Working with the firm they develop a feasibility study including conceptual designs, market studies, permit documents and information needed to decide whether to pursue the opportunity and implement the project. During the pre-construction stage, the developer has fully engaged a development team and an array of outside consultants to get the project ready for construction. The team will undertake value engineering and redesigning to ensure the project is within budget. A major part of this process includes obtaining public approval from governing agencies or departments and making public announcements and presentations to win support of the larger community. Marketing of office space will start early in this stage.

In summary, the three conceptual models of the development process—entrepreneurial, business and real estate—could be condensed into a single conceptual framework but the real estate development process model provides the most suitable framework for this specific study. The real estate development process starts with the idea inception stage and identifies those that will be directly involved in the decision to build green. It also indicates those who are not directly involved in the initiating decision. The real estate development process, however, does not provide details on the financial and non-financial influences that impact or drive the initiating decision to build green.

2.3. Financial Factors

From a contextual place-based perspective, urban core office buildings are often referred to as prime properties based on the value that accrues to them because of their location. Based on geography, the bid rent theory holds that the price and demand for real estate declines as the distance from the CBD increases [28]. This rent gradient is the marginal cost of distance. Simons et al. [16] note that the rent gradient of a property significantly influences the competitive actions of real estate developers. Making decisions about prime property in the CBD rent gradient are high-stakes decisions.

Most studies start with the idea that the major influence on key stakeholders when making the decision to build green will be financial. This focus is evident from the number of empirical studies trying to assess the financial benefits of green buildings. These studies include research in the US [29]; US and Canada [30]; Singapore [31]; and England [32]. Wadu Mesthrige and Chan [15] looked at prime commercial office buildings in Hong Kong used a hedonic-model of rent and building attributes. A US study conducted in 2018 by Fuerst et al. [33] also used a hedonic model. They looked at the price investors paid in sealed bid auctions for foreclosed commercial real estate. Their study analyzed price paid (dependent variable) and various other building attributes including the classification of the building as eco-certified (independent variables). A significant finding was that in the Class A office market segment, eco-certified space (LEED and/or Energy Star) had become part of the Class A office sector and not a niche submarket. LEED building certification had become a factor in the property's market valuation, defined as premium rent [34]. This calculation shapes the investment decisions of developers, investors and owner landlords in determining the degree to that they will pursue specific green building attributes. The question then is, has LEED become a major hedonic factor such as location, design, or age that define a commercial office building as 'investment quality'? [29].

2.4. Non-Financial Factors

There is an equally important body of empirical research that examines various categories of drivers that influence key stakeholders when making the decision to build green. This research stream was developed in 2010 by Falkenbach et al. [35] It is based on a comprehensive review of literature. They compiled a list of drivers that impacted decisions by real estate investors. They identified 10 drivers or benefits from environmentally sustainable buildings. Darko et al. [36] expanded this perspective from real estate investors to a number of stakeholders involved in building construction. This 2017 study identified 64 drivers from a review of 42 selected empirical studies. These were conceptualized in a theoretical framework consisting of five major classifications of green building decision drivers or inhibitors—corporate level, external level, property level, project level, and individual level. Darko et al.'s [36] expansion to include the perspective of more stakeholders follows Zhang's [37] 2015 review of literature. She looked at green development to identify the various stakeholders involved throughout the real estate development process.

Other studies have expanded this investigation of drivers. Whitney et al. [38] looked at the issue of energy usage in Canadian commercial office buildings. Taking a commercial property investor's perspective, a 2020 study by Leskinen et al. [39] reviewed 70 empirical studies that looked at the impact of green certification on discounted cash flow and value. They link these financial factors to property level drivers influencing sustainable building adoption that was suggested by Falkenbach et al. [35] and Darko et al. [36]. A relevant factor that Leskinen et al. [39] looked at is the growing interest in responsible property investing by real estate companies and real estate investment trusts (REIT). These investors are shifting from an asset level approach to a portfolio level approach. In summary, all of the studies that looked at drivers/influencers consider the financial and non-financial factors that shape the decisions of real estate developers, institutional investors and owner landlords.

There are several other conceptual areas of research on decision making that are worth noting. The first is the emerging area of game theory-based analysis of decision making for building or retrofitting green, especially in relation to state provided incentives and/or regulation of developers [40–42]. A second is the expanding area grounded in human behavior theories from psychology and sociology. Another approach deals with ethics and social responsibility discussed in the 2010 research of Eichholtz et al. [43]. They view their study as the first credible evidence on the economic value of a building's certification as green from the perspective of impersonal market transactions rather than engineering estimates. In addition to the direct effects of energy savings, they found that there were intangible effects of green certification that also have a role in determining the market value of green buildings in the marketplace. It is not just rental rates and construction cost per square foot. It can also be 'doing good' at the individual level and corporate social responsibility level. This raises the issue: is it actually 'doing good' or is it self-reporting that they are 'intending to do good'? Research in the field of consumer behavior has long recognized the gap between an individual's intention and behavior [44]. Zhang et al. [45] surveyed the driver mechanisms influencing 156 developers on their action to do redevelopment of industrial brownfields in China. The study concluded that altruistic motives (awareness of responsibility) can close the gap between intention and behavior. An awareness of consequences did not. They reasoned that real estate developers as a group are primarily seeking profit. The fact that green redevelopment could have positive influences on others was not a significant factor to voluntarily engage in green redevelopment.

Contextually, our exploratory study is the first empirical research that examines the impact of LEED certification on property values in Toronto's prime commercial office market. Conceptually, it uses the real estate development process to identify the point that the decision to build green is initiated and the key stakeholders who make this decision. Finally, in a very general way, it seeks to identify financial and non-financial drivers that influence the decision to build or retrofit green. The importance of our research study is that it comes at a critical time for researchers, stakeholders and policy makers if they are to increase efforts to address the issues of global warming and climate change.

3. Decision to Build or Retrofit Green

It is increasingly recognized that real estate development solutions to support SDGs and GHG emission standards will require fundamental changes with regard to how business is conducted. The diversity and creativity in office building architecture suggests an industry open to change. A major part of the pre-construction stage of the real estate development process centers on engagement with government departments and the public suggesting transparency. However, as noted, the decision of whether to build/retrofit green is made by a few stakeholders early in a project's development process [30]. This is supported in the literature [38,39]. The key stakeholders are real estate developers, institutional investors and owner landlords. This suggests that it is a rather closed process.

3.1. *Decision to Use LEED Certification*

LEED certification in Canada is under the Canada Green Building Council (CaGBC). New and major commercial renovations come under LEED Canada NC (new construction) rating system. LEED Canada EB: O&M (existing building: operation and maintenance) provides an entry point into LEED certification for owner landlords of existing buildings [46]. In 2018 Canada placed third in number and square footage of LEED projects worldwide—after the United States and China [12]. Commenting on these efforts Thomas Mueller President and CEO of CaGBC notes:

> 'Canada's building industry is demonstrating how business and sustainability can go hand in hand ... owners and developers are increasingly making LEED an integral part of doing business spurring demand for innovative products ... in the process creating jobs and positive bottom lines' [47].

Will these GB efforts be new office construction or retrofitting of existing buildings? Compared to other global financial centers, Toronto's downtown CBD office space has an extremely low vacancy rate with demand outpacing supply. Downtown Toronto's office vacancy rate fluctuated between one and two percent with an inventory of 91.3 million square feet prior to the Covid-19 pandemic. The asking rent (exclusive of premiums) at the start of the third quarter 2020 was CAD 39.02 per square foot. New office projects that are underway will add 9.4 million square feet. It is estimated that 10 to 15 new projects are anticipated in the next decade [48–50]. The greatest potential for addressing Toronto's growing concerns on GHG emissions, energy-efficiency and operating costs will be created by the greening of existing non-green office buildings and renovating early GBs with older and less efficient green technologies [3–5]. In order for Canada to continue its position in LEED projects, especially in Toronto (Canada's largest commercial real estate market), key stakeholders will need to lead this effort. It is important to understand: the role that key office building stakeholders have in making decisions to pursue a green agenda, and the factors that contribute to their decisions of whether to go green.

3.2. *Decision-Making by Key Stakeholders*

3.2.1. Real Estate Developers

Real estate developers have a key role in the real estate development process. They bring together all of the many stakeholders involved in this process—from idea inception to construction. It is during this process that decisions are made to build or retrofit green. In the absence of a government mandate, real estate developers are the ones who match market demand with the supply of green or conventional office buildings. Although environmental certification protocols such as LEED define, measure and evaluate various levels of green in buildings it is the developer who balances a project's economic, environmental and social aspects. This is evident in the fact that real estate developers begin to market new office space well in advance of construction. Their goal is full occupancy, or a very high percentage of occupancy attained before construction is completed. Mangialardo et al.'s [13] study of office buildings in Milan found that LEED certified properties generated rent premiums and higher prices and they were leased in half the time compared non-green.

When a GB is selected for a new construction project it is critical to first determine which LEED protocols are achievable and incorporate them at an early stage in the project. At that point, these features can be integrated, effectively supporting each other throughout the project [51]. Sustainable buildings and efficiency measures are vital to both developer and tenants but the potential for efficiency and savings is not the same to both. The actual performance of a GB is inevitably connected to the behavior of its occupants—how they actually use energy and resources. These factors can significantly reduce a green building's positive environmental impact [52].

Real estate developers and institutional investors are understandably uncertain about how far to pursue environmental investments because much of the economic rationale for the development of GBs—especially energy costs—is based on site or sub-market specific evidence [29]. It is also difficult for developers to determine what the most cost-effective methods of greening their projects are because buildings can be sources of environmental degradation during their construction, operation, and demolition that requires a life-cycle or cradle-to-grave assessment [53,54].

3.2.2. Institutional Investors

The actions of institutional equity investors in commercial office building development can exercise considerable influence on how sustainable property-related issues are addressed. In 2000, Bartlett and Howard [55] challenged the traditional focus of GB decisions on cost benefit and value methods that indicated GBs cost 5% to 15% more to build. They suggested that decision-makers should consider the whole life cost and environmental impact of buildings. Ten years later, Chegut et al. [56] looked at income performance of the UK's green commercial real estate sector. They reported that GBs received 21% higher rental and 26% higher sales transaction prices both per net square meter, but they added a caveat—that rental contract features (lease term, rent free period, days on the market, etc.) decrease these rental premiums about 5%. Zhang et al.'s [57] "Turning green into gold" offers a comprehensive review of the financial side of 'going green'. They distinguish between profitability from a building life cycle perspective by major market participants and economic viability perspective by developers and occupants. Hsieh et al. [58] shifts the discussion from owner's benefits to an examination of capital markets including institutional investors and building green. Their findings indicate that the cost of equity capital for the development of LEED certification green buildings was lower. Prior studies have reported higher cost of equity capital during the development period [59].

While there is no question that financial considerations play a primary role in the decisions of institutional investors, it is also true that over the past twenty years non-economic consideration have gained a place in the equation and in corporate public reporting [60]. The institutional investors partnering with real estate developers in commercial office buildings in Toronto include pension, endowment and mutual funds; insurance companies; and commercial banks. All of them have had to adjust to corporate social responsibility policies; socially responsible investment; and environmental, social and governance (ESG) sustainable investing. There is a direct link between investor's financial metrics, the efficiency and productivity of building projects and achieving the 2030 United Nations Agenda and its 17 SDGs [9]. Globally, society is expecting the public and private sectors to take the lead in adopting sustainable practices that address critical ESG issues.

Nevertheless, while some studies have found evidence on social and corporate responsibility with no strings attached—simply doing good [43]—much of the evidence points to financial motives [51]. In order to persuade developers, investors, and owners of the benefits of eco-investments, the payoff from investment in green buildings needs to be identified. Early studies found the strongest drivers of responsible property investing to be traditional considerations such as opportunities to outperform the market [61]. Evidence continues to emerge that LEED labelled buildings achieve a higher financial return than conventional buildings in terms of rental and sales prices. Overall, a green premium is considered to be a potential driver of investment in labelled buildings [62]. To date, however, it is still unclear to what extent the adoption of green practices and expectation for financial performance are changing business models from traditional to sustainable, especially when applied to specific geographic markets [63].

3.2.3. Owner Landlords

Many commercial building owners are single occupant firms that contract for the construction of an office building on land they have purchased. This type of owner constructs a LEED certifiable building to house its employees and bear its corporate name. In this study the focus is on a separate category of owners—owner landlords—that seek an equity investment in an office building producing rental income from commercial tenants and leaseholders. Qui et al.'s [64] study of commercial buildings in the state of New York found that if the buildings were owner occupied there was less likelihood of green certification. Real estate developers and owner landlords are very often one and the same, but owner landlords can also be institutional investors such as pension funds that participate in the building's entire life cycle. Once constructed they usually contract with a property management firm or real estate agency to handle marketing, tenant relations and manage building operations.

LEED building characteristics are a factor in property market valuation (premium rents) and the operating cost for owner landlords. These calculations shape the investment decisions of developers, investors and owner landlords in determining the degree that they will pursue specific green building attributes. The main traditional hedonic factors that define a commercial office building as 'investment quality' are: location; condition; design; quantity and quality of floor space; amenities and service; adaptability to the requirements of a tenant; and infrastructure, proximity to transportation and communications [29]. In assessing the contributing value of these attributes, professional property appraisers such as the Royal Institute of Chartered Surveyors will search for and analyze comparable market data. Owner landlord decisions based on sustainability factors will focus on the 'green benefits' that accrue over the life of the building. Therefore, in the real estate development process, owner landlords will primarily be interested in the GB cost/benefit analysis [65] pre-construction phase or architect's design phase [66]. Once the developer hires an architect and consultants on sustainability, engineering, geotechnical, landscape, interior, general contractor, legal, etc., the work can begin. Operationally, building information modelling (BIM) tools are used to assess the buildings environmental sustainability that can include a life cycle assessment (LCA) and lifecycle costing (LCC) analysis [67].

Beginning in the pre-construction phase the developer will begin to market space in the new building, providing the first 'market test'. Tenants that choose to lease/rent space in a LEED certified office building usually sign a triple net lease. The tenants receive the direct benefits from operating cost reduction in LEED certified office space. The developers, investors or owner landlords receive the benefit of higher lease rates paid by tenants [68]. If tenants are willing to pay a higher rate it is an indication that the market is responding favorably to LEED certified office space. When determining to what extent investors and developers should bear additional costs—in efforts to design and construct more socially responsible property—it is important to keep in mind that market supply and demand factors dictate the rental level. Ultimately it is the business productivity of tenants that dictates the tenant's ability to pay [69]. For investors, this means that it is important to consider the economic impact that green buildings have on tenants, including economic benefits such as reduced employee health care costs and sick leave as well as increased performance of employees [13].

Research conducted at Carnegie Mellon University for the US government found that costs associated with employees amounted to 78 percent of total costs. Costs linked directly to the building—rent, operations, maintenance, and office—made up only 9 percent [70]. Many of the cost/benefit studies conducted in the past have used these figures in attempts to estimate the economic gains that can be attained from improvements in the indoor environment offered by green buildings. Even with the best information available, there is still a high level of uncertainty with estimates on health and associated economic gains from improvements in the indoor environment. The largest source of uncertainty is the degree that these health effects could be reduced through practical changes in building design, operation, and maintenance [71]. To date, indoor environment quality of green offices studies report occupants' positive responses but have a lack of evaluating research on the performance and identification of individual indoor environment quality factors [72]. Such research would help tenants decide if they wanted to rent in a green building and would guide developers and owners in the decision to build green.

4. Research Methodology

4.1. Study Objectives

This exploratory study has two objectives. The first is to determine the impact of LEED certification on the market value (expressed as asking rent) for Toronto's CBD urban core office buildings. Studies have found that that there is a premium for LEED-labeled buildings both in terms of rent and sales price [73,74]. The literature suggests that green certification creates a market premium for green labeled buildings. This in turn increases green certification adoption and green construction and retrofitting by key stakeholder decision makers [62]. In accordance with the market and dynamics of Toronto's office building real estate, the existence of green rent premiums could potentially drive the investment and development of new LEED-labeled buildings, as well as the green-retrofitting of the existing uncertified buildings [75].

The second objective of this study is to identify the factors that influence key stakeholders—real estate developers, institutional investors, and owner landlords—with regard to whether or not to seek LEED certification for their office buildings. This initial decision takes place during the idea inception stage of the real estate development process [17]. It is primarily an internal decision with only a few executive level personnel participating in the decision [27]. Both quantitative [13] and qualitative [39] research studies suggest that the decisions by most stakeholders are driven by financial factors. Research studies that examine corporate sustainability performance such as meeting 2030 UN SDGs or ESG targeted-REITs, however, indicate that non-financial factors are the primary drivers during the decision-making process [42].

Based on these objectives, two hypotheses are proposed. Hypothesis (H_1): LEED certification will have a significant positive factor in the asking rent (market value) of office buildings in Toronto's CBD urban core. Hypothesis (H_2): Financial drivers will be more influential than non-financial drivers in a key stakeholder's decision to pursue LEED certification.

The study's research design uses two stages. The first stage was designed to obtain quantitative data for a statistical analysis of specific tangible building variables in relationship to market asking rent valuation to test Hypothesis (H_1). The statistical data provides an analysis, answer and insight on the first hypothesis. The second stage consisted of interviews with experts in Toronto's real estate development sector. It provided qualitative data to test Hypothesis (H_2). The interview data offers clarification, explanation and answers to many of the issues addressed in this study.

4.2. Stage One Reserch Design

Empirical studies continue to find that most decisions on the adoption of LEED certification are primarily influenced by financial implications. Traditionally one of the major considerations is market valuation, expressed as the asking rent of office space per square foot. This study uses an approach used by similar studies that were conducted in Milan [14] and Hong Kong [16]. These studies use a hedonic model in the form of an ordinary least squares (OLS) linear regression model based on a fixed effects approach to the attributes of commercial office property on rental price.

The independent variables that have been shown to be significant determinants on the dependent variable asking rent are presented in Table 1.

Table 1. Studies of hedonic factors impacting office rents.

Author	Region	Sample Size	Dependent Variable	Independent Variables Found to Be Significant
Clapp [76]	Los Angeles metropolitan area	105	Average 1974 asking rent	size, age, number of floors, internal parking, prestigious address, property tax, air quality, amount of office space within a two-block radius, distance by road to nearest motorway junction, average community time for employees
Hough and Kratz 1983 [77]	Chicago central business district	139	Average 1978 asking rent	existence of 'good' architecture, distance from CBD, public parking, age, size, number of floors, availability of conference facility
Cannaday and Kang 1984 [78]	Champaign- Urbana, Illinois	24	Average 1979–1980 asking rent	age, minimum lease term in years, crow fly distance to the CBD, crow fly distance to a shopping centre, average unit size, average number of units per floor
Brennan, Cannaday, and Colwell 1984 [79]	Chicago central business district	29	Actual transacted lease values (incorporating lease terms) within a building from 1980–1983	size of building, size of each unit, lease terms, loss factor (proportion of area rented but not possible to use), position within the building, location with respect to centre of CBD
Glascock, Jahanian, and Sirmans 1990 [80]	Baton Rouge, Louisiana	675	Asking rents of office units from 1985–1988	location, building type, size, the year in which the property was let
Mills 1992 [81]	Chicago	543	Asking rents and the discounted rent over the period of a 15-year lease	age, size, parking, internal restaurant, internal bank, location outside the CBD (but not subsectors within the CBD)
Dunse and Jones 1998 [82]	Glascow, Scotland	477	Asking rents 1994–1995	size, age, location, air conditioning, acoustic tiling, carpeting cellular layout, double glazing, internal parking, raised floors, tea preparation area

Building location (theory of rent gradient [28]), age and size were the variables found to most consistently explain the variation in the dependent variable asking rent. Each study shows that a variety of building attributes related to rent are significant, but they appear significant on a less consistent basis than location, age and size, possibly because the value attributed to them is unique to the particular office market studied. Although previous studies provide insight on which combinations of variables have been proven significant in other markets the results are not necessarily transferable to the commercial office market in Toronto. Based on a review of the independent variables, in these studies we generated a list of principal determinants of rent for local market areas that could then be used to construct a model for downtown Toronto. Data from a quantitative analysis will be used to determine the variance in office rents and whether LEED certification is a significant independent variable accounting for variance in office rents in the CBD and northern business district of Toronto.

4.2.1. Data Collection, Sample and Variables

Data Collection. Information related to specific building variables for both LEED certified buildings and comparable non-LEED buildings was obtained from the Altus Group Altus InSite database, Canada Green Building Council and Toronto City Hall's property assessment database. The Toronto office market in 2016 included 1293 office buildings with about 160 million square feet of space, nearly three quarters of that is in the CBD. In addition, Toronto's downtown core is one that has the highest number of LEED certified buildings in Canada. This provided a sufficient sample size making it an ideal site for investigation [83].

Study Sample. Toronto's main office market has a total office inventory of 75.7 million square feet. The average gross rent is CAD 42.35 per square foot [84]. Gross rent is the monthly rent charged to occupy an office space, calculated to include all operating costs (i.e., maintenance, taxes, utilities, etc.). Two geographic areas of Toronto's main office building market were selected for this study—the Downtown CBD and Northern Toronto (north part of the old Toronto district). Both of the areas were used. The comparison of two submarkets will provide for an accounting of various location characteristics that may impact the value of LEED in any given submarket but are not explicitly accounted for in this study [34]. CBD urban core office buildings in this study will refer to the area bounded by Bloor Street to the North, Lake Ontario to the south, the Don River to the East, and Bathurst Street to the West. Northern Toronto office buildings in this study fall outside of the CBD urban core. They are approximately 13 km North of 200 Bay Street and are within a 3.5 km radius of one another. For purposes of this study, 200 Bay Street is considered to be the center or rent gradient theory point with the most prestigious prime office area in Toronto. Additionally, when choosing the comparable office buildings for the study, only those properties selected were ones that had an area in excess of 30,000 square feet and had an owner landlord but were not owner-occupied (corporate headquarters or offices).

Initially a sample of 114 office buildings was obtained—24 LEED certified (Canada Green Building Council database) and 90 non-LEED comparable properties (Toronto City Hall's building database). The final sample—after being reduced because of study specifications and limitations related to data availability—provided 16 LEED certified buildings and 52 non-LEED certified comparable buildings. Geographically, the sample included 21 buildings from Northern Toronto (four LEED certified and 17 comparable buildings) and 47 buildings from Toronto's CBD (12 LEED certified and 35 comparable buildings). For each LEED building included in the study attempts were made to include three or more non-LEED certified but otherwise comparable office buildings (see Table 2).

Table 2. Sample of Toronto commercial office buildings.

Location	LEED Certified	Non-LEED	Totals
CBD Toronto	12	35	47
Northern Toronto	4	17	21
Totals	16	52	68

Study Variables. Eleven independent variables were selected and used in this study. Although this list is not exhaustive, the selection is consistent with the precedents of prior research (see Table 1). These eleven provide a comprehensive list of variables (independent variables) to be used in a hedonic regression analysis model in which the attribute of asking rent per square foot of an office space in a commercial office building is the dependent variable. This study uses asking rent as a proxy or measure for current market value. Attempts were made to obtain data related to actual transacted rents with a number of owner landlords, real estate brokers and research firms but those were unsuccessful—most did not want to disclose this information. Although concerns can be raised on the issue of using asking rents rather than actual transaction price this has been used in most studies. Dunse and Jones's [82] study of the office rental market in the city of Glasgow found initial asking rent and final transacted rent to have a correlation coefficient of 0.98, showing that the two variables are highly correlated. Oyedokun et. al. [34] cautioned that the academic hedonic regression models dealing with green rent premium studies need to guard against three things: missing variables; coefficients based on equilibrium assumptions during periods of market volatility; and distinguishing between a new and a green office premium when green offices represent a high proportion of new offices in a study.

Eleven building measurement factors were selected as independent variables. They were divided into five categories of attributes (see Table 3). The first is the attribute of green certification—indicated by whether a building is LEED certified. The second category focuses on a building's physical and structural attributes. This category included: age; total building office area, parking stalls per 1000 square feet of leasable office space; and BOMA metropolitan base definitions for three building Classes: A, B, or C [85]. The third category is vacant office space at the time of the study—indicated by the direct percentage of building spaces available for lease from the landlord or as a sublease from existing tenants. The fourth category is based on a building's common area size such as lounges, atrium, hallways, etc. It is measured by the total additional monthly rent per square foot that is charged to tenants above asking rent. This charge is the landlord's costs for maintaining these areas. Total additional rent is also included due to its impact on the gross rent that tenants pay. The fifth category that impacts the variation in rent is the office building's location (theory of rent gradient [28]). Due to the clustering of buildings in this study it was determined to calculate the distance of each building from Toronto's CBD center point (200 Bay Street). This was used to define location. This Euclidian straight-line 'as the crow flies' distance approach was used to account for any variation in office rents related to building location—variance usually associated with a prestigious address. The address of 200 Bay Street is at the heart of Toronto's business district, an area predominantly occupied by financial institutions and large professional practices. It also commands top rents for office space, providing justification for its use as the epicenter of the CBD in this study.

Table 3. Description of property attributes in the model.

Location	Variables and Abbreviation	Measure	Description	Expected LEED Correlation
Rental price	NET RENT	Continuous	Asking rent per square foot	+
Green certification	LEED	Dummy	LEED certified	1.0
Building's physical structure	CLASS A	Dummy	Competes for premium users	+
	CLASS B	Dummy	Wide range of users, does not compete with Class A	-
	CLASS C	Dummy	Tenants require functional space	-
	OFFICE AREA	Continuous	Total building square footage	+
	PARKING	Continuous	Parking spots per 1000 square feet of leasable space	+
	AGE	Continuous	Age of building in years	-
Vacant office space	DIRECT	Continuous	% of building space available for lease from landlord at time of study	-
	SUBLET AVAILABLE	Continuous	% of building space available for lease from an existing tenant at time of study	-
Common area size	ADDITIONAL RENT	Continuous	Total additional charges (per square foot) over asking rent	+
Location	CROW FLY	Continuous	Euclidian distance from subject property to 200 Bay Street	-

4.2.2. Data Analysis

Data gathered in stage one was subjected to two techniques of statistical analysis. First, the generation of bivariate Pearson correlation coefficients (Pearson's r) to statistically measure the strength and linear direction in paired dyadic relationships for all of the variables. A correlation matrix was produced from this paired relationship analysis revealing a range of values from strong negative (−0.931) relationships to strong positive (+0.650) relationships. The statistical significance of the correlation coefficients (r values) was measured by a two tailed test probability distribution for p-values at the 0.05 and 0.01 levels.

Second, in keeping with the approach used in a number of studies [86–89], an ordinary least squares (OLS) linear regression analysis was used. This analysis was used to develop a hedonic model that statistically accounts for the variance in office asking rents (the regressand) in relation to eleven office building attributes. Specifically, these findings will test Hypothesis (H_1)—LEED certification will be a significant positive factor in the asking rent (market value) of office buildings in Toronto's CBD urban core.

4.3. Stage Two Research Design

After quantitative data was collected in the first stage of the research, qualitative data was obtained in the second stage. The method of collecting this information was one-on-one interviews with three executive-level experts. This was to obtain first-hand information on Toronto's office building sector, where it is on building green and where it is headed. This method of qualitative empirical research has wide acceptance in social science research [90]. Experts are considered knowledgeable on a particular subject. They are identified as experts by virtue of their knowledge, position, and status. As key informants, experts provide exclusive knowledge, relevant experience, and an executive perspective on the complexity of decision making involved in adoption of green initiatives such as LEED certification [88].

Several of Toronto's large and influential commercial real estate development firms were contacted and three executives agreed to be interviewed. Interviewee's statements were recorded, later transcribed for analysis and are included in the Appendix A. The credentials of the three executive-level experts were outstanding.

Interviewee 1: serves as Vice President for one of Canada's largest landlord developers, and in that role is responsible for the origination and execution of office, industrial, and land transactions, together with major property portfolios across Canada.

Interviewee 2: serves as Senior Vice President for one of Canada's largest landlord developers, and in that role is involved in informing the decision-making that provides the company with strategic direction.

Interviewee 3: serves as the National Director of Sustainability and Energy Management for one of Canada's largest landlord developers, and in that role informs company decision-making related to making the company more environmentally responsible and energy efficient.

A semi-structured expert interview approach was used to ensure that each interviewee was presented with the same questions; this was to increase the reliability and credibility of the data. The interviews focused on understanding the financial and non-financial factors that key stakeholder firms consider with regard to LEED certification during construction of new office buildings or retrofitting existing ones. Probing questions were asked as follow-on questions to obtain more in-depth information, especially if answers seemed insufficient in detail or more clarification to answers was needed. Special attention was given to reasons they see for changing their focus to green and assessing the impact that LEED has on market value of their properties. The interview data provides clarifications, explanations and answers for the second hypothesis: Hypothesis (H_2)—Financial drivers will be more influential than non-financial drivers in a key stakeholder's decision whether to pursue LEED certification. Analysis of the data, especially with a limited number of interviews, includes a summary by topics and ideas rather than a statistical analysis.

5. Stage One Results and Discussion

5.1. Findings from Correlation Coefficient Matrix

Our first step of analysis was to construct a correlation matrix to determine the explanatory power of each variable (see Table 4). The results of the correlation matrix indicate that the correlation of LEED certification with net asking rent is not significant at the 0.05> level. This finding does not support Hypothesis (H_1) = LEED certification will have a significant positive factor in the asking rent (market value) of office buildings in Toronto's CBD urban core. The impact of LEED certification on the market value (asking NET RENT) of an office building was not found to be statistically significant. Although the correlation of 0.193 was found not to be statistically significant at the 0.05> level a review of empirical studies suggests several possible interpretations. First, as Fuerst et al. [33] and Oyedokun et al. [34] suggest, office buildings with LEED certification generally have other attributes sought by premium office space renters, thus having a higher correlation with NET RENT. The 0.01> level positive statistically significant correlation between NET RENT and ADDITIONAL RENT (0.873 **), PARKING (0.616 **), OFFICE AREA (0.566 **) and CLASS A (0.489 **) are all quality building attributes that can demand higher rent per square foot. The high negative correlations between NET RENT and several other attributes support this conjecture about relationship to building quality. Negatively correlated with NET RENT are CROW FLY location (−0.608 **), CLASS B (−0.428 **), and CLASS C (−0.198). While Class A are prestigious buildings and compete for premier office users, Class B buildings are fair to good in attributes and location; and Class C buildings are below average for the area and compete for tenants seeking functional space [85].

Table 4. Correlation matrix.

	Net Rent	LEED	Class A	Class B	Class C	Office Area	Parking	Age	Direct	Sublet Available	Additional Rent	Crow Fly
NET RENT	1	0.193	0.489 **	−0.428 **	−0.198	0.566 **	0.616 **	0.089	0.214	−0.032	0.873 **	−0.608 **
LEED		1	0.394 **	−0.367 **	−0.102	0.230	0.190	−0.197	−0.060	0.025	0.240	−0.104
CLASS A			1	−0.931 **	−0.260 *	0.390 **	0.368 **	−0.196	0.001	−0.069	0.489 **	−0.110
CLASS B				1	−0.110	−0.348 **	−0.303 *	0.185	0.008	0.019	−0.406 **	0.081
CLASS C					1	−0.144	−0.201	0.042	−0.022	0.137	−0.256 *	0.085
OFFICE AREA						1	0.550 **	−0.089	0.198	−0.063	0.650 **	−0.297 *
PARKING							1	0.113	0.135	−0.230	0.649 **	−0.455 **
AGE								1	−0.109	0.058	0.031	−0.171
DIRECT									1	−0.038	0.172	0.060
SUBLET AVAILABLE										1	0.015	−0.152
ADDITIONAL RENT											1	−0.576 **
CROW FLY												1

** Correlation is significant at the 0.01 level (two-tailed). * Correlation is significant at the 0.05 level (two-tailed).

Next, consider the correlations between these building attributes and LEED certification. The only two attributes that LEED is significantly correlated at the 0.01 > level are positive with CLASS A (0.396 **) and negative with CLASS B (−0.367 **). The other attributes that LEED certification has high ratings that offer some insight. There is positive direction of relationship between LEED with ADDITIONAL RENT (0.240), OFFICE AREA (0.230) and PARKING (0.190). These three attributes relate to building area and are positively correlated with LEED because the buildings in the sample that were LEED certified tend to be larger buildings. It follows that larger office buildings are also those that are more likely to have a large amount of common area, parking and more amenities thereby increasing the amount of additional rent charged to tenants. There is negative direction of relation between LEED with AGE (−0.197), location CROW FLY (−0.104), CLASS C (−0.102) and DIRECT from landlord lease space available. The negative correlation coefficients with AGE, CROW FLY and DIRECT suggest that LEED buildings tend to be newer properties, situated near the center of the CBD, and have lower than average vacancy.

These negative correlations would suggest an interpretation that LEED certification is a strong positive factor in terms of its position among office building attributes as well as in relation to NET RENT. Then why doesn't LEED certification correlate significantly with NET RENT? Based on their UK study on the growth of green office buildings, Oyedokun et al. [34] would propose that green certification is becoming mainstream. Fuerst et al.'s [33] study of the US office market might go as far as to propose that LEED certified space is becoming a defining factor in Class A. Three points need to be stated with respect to this study. First, all LEED certified buildings included in the sample are Class A buildings thereby resulting in a strong positive correlation with CLASS A and a strong negative correlation with CLASS B. Furthermore, there were far fewer Class C buildings included in the study than Class A and Class B buildings. Second, this is a place-based study focused on Toronto's urban downtown core and not the Greater Toronto Area—the most populous metropolitan area in Canada. It would be expected that the office building sector would be dominated by Class A buildings. Third, the low level of significance for these variables based on the correlations makes it difficult to generalize about these beyond the study sample. However, as an exploratory study it provides the groundwork for further research.

5.2. Findings from Linear Regression Model

Once the correlation relationship (see Table 4) between an office building's LEED certification and the coefficients of the other building attributes were analyzed, the hedonic linear regression model was constructed. An ordinary least squares (OLS) analysis was conducted to determine which compilation of variables created the equation that best accounted for the variance in net asking rents (NET RENT). In the analysis, the independent variable with the highest partial correlation coefficient to NET RENT is entered first into the model. This process is continued by adding the variable with the next highest partial correlation coefficient to the model and the adjusted R^2 for the new model is evaluated to compare its explanatory power to that of the previous model. This process is continued until all of those variables with a partial coefficient significant at the 0.05 level have been added to the model to assess their impact on the model's explanatory power. In addition, as new variables are added to the model, previous variables are removed from the equation if their significance level falls below the 10 percent critical value.

The first variable included in the model was ADDITION RENT which had the greatest positive partial correlation coefficient of 0.873 ** with NET RENT. The initial model returned an adjusted R^2 of 0.757, with ADDITIONAL RENT significant at the 99 percent critical value level. This high adjusted R^2 may be cause for concern that 'ADDITIONAL RENT' might be drowning out the effect that other variables (including LEED) might have on NET RENT, especially when considering the degree of correlation between ADDITIONAL RENT and the other predictor variables. The process of adding and removing variables from the model based on their partial correlation coefficients and significance levels was continued until arriving at a model consisting of ADDITIONAL RENT and CLASS B. The model produced an adjusted R^2 of 0.769, with ADDITIONAL RENT significant at the 99 percent critical value level and CLASS B significant at the 95 percent critical value level (see Table 5a–c). We describe the linear regression model as our 'best model'.

Table 5. (**a**) Linear regression model summary; (**b**) Linear regression ANOVA; (**c**). Linear regression Coefficients.

(a) Linear regression model summary.

Model Summary

Model	**R**	**R Square**	**Adjusted R Square**	**Std. Error of the Estimate**
CLASS B ADDITIONAL RENT	0.881 [1]	0.776	0.769	3.59315

(b) Linear regression ANOVA

Model	**Sum of Squares**	**df**	**Mean Square**	**F**	**Sig.**
Regression	2772.972	2	1386.486	107.390	0.000 [3]
Residual	800.464	62	12.911		
Total	3573.437	64			

(c).Linear regression Coefficients.

Coefficients [4]

Model	**Unstandardised Coefficients**		**Standardised Coefficients**	**t**	**Sig.**
	B	**Std. Error**	**Beta**		
Constant	−5.585	2.340		−2.387	0.020
ADDITIONAL RENT	1.277	0.100	0.100	12.808	0.000
CLASS B	−2.195	1.087	1.087	−2.018	0.048

[1] Predictors: (Constant), CLASS B, ADDITIONAL RENT. [2] Dependent Variable: NET RENT. [3] Predictors: (Constant), CLASS B, ADDITIONAL RENT. [4] Dependent Variable: NET RENT.

The exclusion of variables from the best model can be explained when we examine the relation of each variable to ADDITIONAL RENT while taking into consideration the characteristics of the sample buildings. The first variable that was removed from the model was PARKING which has a partial correlation coefficient of 0.616 ** but could still not be deemed significant in a model that already included ADDITIONAL RENT. This is because ADDITIONAL RENT is typically a composition of those costs associated with property taxes, common area maintenance, and any other additional expenses the owner may charge related to building maintenance and services. The greater the amount of common area a building possesses, the higher the ADDITIONAL RENT. Those buildings with the greatest amount of common area in downtown Toronto are located in the CBD, an area where reserved parking is rare and highly sought after. When linking the value placed on PARKING space in the CBD with the higher ADDITIONAL RENT in the area it is understandable why the significance of PARKING was drowned out by the ADDITIONAL RENT variable.

Going down the list of variables with high individual correlation coefficients we see that the same effect was had on CROW FLY location and total building OFFICE AREA. These variables too are affected by proximity to the centre point of the CBD. CROW FLY is affected because it is actually a measure of building proximity to the centre of the CBD while OFFICE AREA is affected because the buildings in downtown Toronto with the greatest amount of office space are those in the centre of the city's CBD. The only other variable excluded from the model with a significant individual correlation to net asking rent was CLASS A. CLASS A was excluded because it was replaced by CLASS B that reduced its significance and provided a model with a higher adjusted R^2. The large changes in the regression coefficients produced by the inclusion of ADDITIONAL RENT in the model are an indication of multicollinearity. The tendency toward redundancy as noted is due to the nature of the model that included multiple factors that respond not only to the response variable NET RENT but also to each other. This explains the relationship between CLASS A and CLASS B although the multicollinearity that exists between these two variables was expected because both are indicators of building class. For this study we chose to evaluate both to see which contributed most significantly to the model.

In an attempt to reduce multicollinearity, a second stepwise regression analysis was performed that excluded ADDITIONAL RENT from the model. The first variable included in the new model was PARKING (partial correlation coefficient of 0.616 **), resulting in an adjusted R^2 of 0.370 with PARKING significant at the 99 percent critical value level. CROW FLY was added to the model next returning an adjusted R^2 of 0.501 while both coefficients maintained significance at the 99 percent critical value level. OFFICE AREA was then included, increasing the adjusted R^2 to 0.552. Although CROW FLY and OFFICE AREA remained significant at the 99 percent critical value level the significance of PARKING was reduced to the 95 percent critical value level.

When CLASS A is added to the model, we see the significance of PARKING and OFFICE AREA were reduced to the 90 and 95 percent critical value levels respectively while CROW FLY and CLASS A are significant at the 99 percent critical value levels and the model's adjusted R^2 moves to 0.604.

Finally, CLASS B is introduced to the model but subsequently removed due to its negative effect on the significance of the other variables and adjusted R^2. This means that our best model is comprised of the coefficients PARKING, CROW FLY, OFFICE AREA and CLASS A, producing an adjusted R^2 of 0.604 (see Table 6a–c).

Table 6. (**a**) linear regression summary; (**b**) Linear regression ANOVA; (**c**) Linear regression coefficients.

(a) linear regression model summary				
Model Summary				
Model	**R**	**R Square**	**Adjusted R Square**	**Std. Error of the Estimate**
PARKING	0.793 [1]	0.629	0.604	4.70200
CROW FLY				
OFFICE AREA				
CLASS A				

(b) Linear regression ANOVA					
ANOVA [2]					
Model	**Sum of Squares**	**df**	**Mean Square**	**F**	**Sig.**
Regression	2246.908	4	561.727	25.407	0.000 [3]
Residual	1326.528	60	22.109		
Total	3573.437	64			

(c) Linear regression coefficients					
Coefficients [4]					
Model	**Unstandardised Coefficients**		**Standardised Coefficients**	**t**	**Sig.**
	B	**Std. Error**	**Beta**		
Constant	17.839	1.404		12.710	0.000
PARKING	0.001	0.001	0.200	1.907	0.061
CROW FLY	0.000	0.000	−0.409	−4.616	0.000
OFFICE AREA	0.000	0.000	0.226	2.324	0.024
CLASS A	4.311	1.432	0.264	3.010	0.004

[1] Predictors: (Constant), PARKING, CROW FLY, OFFICE AREA, CLASS A. [2] Dependent Variable: NET RENT. [3] Predictors: (Constant), CLASS B, ADDITIONAL RENT. [4] Dependent Variable: NET RENT.

According to the correlation matrix the correlation of LEED with net asking rent is not significant at the 0.05 level. This would normally disqualify it from inclusion in the model, but regressions were still conducted to assess the impact of LEED on the explanatory power of our best model. Ultimately, when LEED was introduced to the model all original model coefficients fell within their original significance levels, but LEED was not shown to be statistically significant (0.528), pulling adjusted R^2 down to 0.600. It was therefore concluded that LEED certification should remain excluded because it did not improve the explanatory power of the model.

When CLASS A is added to the model the significance of PARKING and OFFICE AREA are reduced to the 90 and 95 percent critical value levels respectively while CROW FLY and CLASS A are significant at the 99 percent critical value levels and the model's adjusted R^2 moves to 0.604.

As previously noted, the Pearson correlation coefficient matrix (Table 4) reveals that the correlation of LEED certification with NET RENT is not significant at the 0.05> level which causes Hypothesis (H_1) to be rejected. This would normally disqualify LEED from inclusion in the model, but regressions were still conducted to assess the impact of LEED on the explanatory power of our best model. Ultimately, when LEED was introduced to the model all original model coefficients fell within their original significance levels, but

LEED was not shown to be statistically significant (0.528), pulling adjusted R^2 down to 0.600. It was therefore concluded that LEED certification should remain excluded, because it did not improve the explanatory power of the model with regard to the NET RENT (i.e., asking rent as market value) of green buildings (see Table 7a–c).

Table 7. (**a**) linear regrèssion model summary; (**b**) Linear regression ANOVA; (**c**) Linear regression coefficients.

(a) linear regression model summary

Model Summary

Model	**R**	**R Square**	**Adjusted R Square**	**Std. Error of the Estimate**
LEED CROW FLY OFFICE AREA CLASS A PARKING	0.795 [1]	0.631	0.600	4.72555

(b) Linear regression ANOVA

ANOVA [2]

Model	**Sum of Squares**	**df**	**Mean Square**	**F**	**Sig.**
Regression	2255.920	5	451.184	20.205	0.000 [3]
Residual	1317.517	59	22.331		
Total	3573.437	64			

(c) Linear regression coefficients

Coefficients [4]

Model	**Unstandardised Coefficients**		**Standardised Coefficients**	**t**	**Sig.**
	B	**Std. Error**	**Beta**		
Constant	17.839	1.411		12.651	0.000
PARKING	0.001	0.001	0.200	1.897	0.063
CROW FLY	0.000	0.000	−0.411	−4.615	0.000
OFFICE AREA	0.000	0.000	0.231	2.351	0.022
CLASS A	4.620	1.520	0.283	3.040	0.004
LEED	−0.922	1.451	−0.055	−0.635	0.528

[1] Predictors: (Constant), LEED, CROW FLY, OFFICE AREA, CLASS A, PARKING. [2] Dependent Variable: NET RENT. [3] Predictors: (Constant), LEED, CROW FLY, OFFICE AREA, CLASS A, PARKING. [4] Dependent Variable: NET RENT.

6. Stage Two Research Results and Discussion

Stage Two was designed to obtain information on the issues examined in this study relative to the development of office buildings in Toronto. Three executives in key stakeholder firms in Toronto were interviewed using a semi-structured format and a five-question protocol with follow-on questions to ensure that their answers were complete and explanatory. These interviews were recorded and transcribed for analysis; the full transcripts are included in the Appendix A.

Reports on LEED certification show that the number of green office buildings in Toronto have increased during the past twenty years [8,83]. Using an interview format rather than a written questionnaire provides an opportunity to obtain first-hand knowledge on the role of key stakeholders in decisions related to building or retrofitting green and the factors that impact their decisions. Answers to the questions posed to the expert interviewees provide valuable insight related to the second hypothesis: Hypothesis (H_2)

= Financial drivers will be more influential than non-financial drivers in the decision of whether to pursue LEED certification.

A brief synopsis of the three expert's qualifications is provided.

Interviewee 1: Serves as Vice President for one of Canada's largest landlord developers. In that role is responsible for the origination and execution of office, industrial, and land transactions, together with major property portfolios across Canada.

Interviewee 2: Serves as Senior Vice President for one of Canada's largest landlord developers. In that role is involved in informing the decision-making that provides the company with strategic direction.

Interviewee 3: Serves as the National Director of Sustainability and Energy Management for one of Canada's largest landlord developers. In that role informs company decision-making related to making the company more environmentally responsible and energy efficient.

6.1. *Findings from Interviews*

Each interviewee was asked the same questions, although the follow on probing questions changed with each interview. Their insights are valuable, so readers are encouraged to examine the full transcripts in the Appendix A. Table 8 provides a summary of interviewee responses to the five-question protocol used.

Table 8. Summary of interviewee responses.

Questions	Interviewee 1	Interviewee 2	Interviewee 3
Does LEED certification currently have a significant impact on achievable gross rent?	No	No	No
Moving forward, do you think LEED is going to be the new benchmark (i.e., the new Class A)?	No	Yes	Yes
Could a building built in compliance with LEED, but without certification (label), achieve the same returns as a LEED certified (labelled) building?	Yes	Yes	Yes
Will LEED certification help attract more potential purchasers/tenants?	Yes	Yes	Yes
Do you plan on retrofitting existing buildings to LEED standards?	No	Yes	Selectively

6.1.1. Question 1: Does LEED Certification Impact Achievable Gross Rent?

Interviewees were asked if LEED certifications currently have 'a significant impact on achievable gross rent'. The question refers to achievable gross rent rather than net asking rent (the dependent variable in this study) because, as noted in the research [91], we wanted to distinguish the impact of LEED certification from factors that impact those of operating costs on rents. LEED buildings tend to have lower operating costs so it is assumed that owners will increase the net rent portion of the formula to keep their properties comparable to the market on a gross rent basis [92]. Although potential savings in operating expenses allows for possible increases in net asking rents the expert interviewees did not see LEED as having a significant impact on achievable gross rent. All interviewees noted that the lead tenants, institutional investors (especially pension funds) and occupants linked to firms with corporate social responsibility policies wanted sustainability green platforms in buildings. This agrees with the finds of Eichholtz, et al. [43].

6.1.2. Question 2: Is LEED Going to Be the New Benchmark (New Class A)?

Office buildings in Canada are subjectively classified and ranked in descending order A, B., or C [85]. As LEED certified buildings already carry a premium asking rent, we wondered whether LEED is going to be the new benchmark (new Class A)? Interviewees 2 and 3 said yes, they felt it would. Interviewee 1 differed from the others in response to this question. The reason for the difference of opinion was that Interviewee 1 approached the question from the standpoint of what tenants' value which is gross rent. If LEED is not considered to have a significant impact on gross rent it is not likely to become a factor in developing a new building standard. Interviewee 2 approached the question from the standpoint of what valuation professionals consider during the property appraisal process, noting that there is a factor built into appraisal methodology that gives green buildings more value because they last longer. Finally, the Interviewee 3 noted that when they first decided to pursue LEED certification it was a marketing advantage (strategy) while now it is becoming the accepted standard. Based on a review of the literature, the interviews and LEED certification data indicate that LEED is becoming part of new office buildings standards although it is not yet the new benchmark [33,34]. Perhaps a more important question is, who are the drivers for the focus on green office buildings? Are the owner landlords, real estate developers and valuation appraisal professionals creating or following the market? Are institutional investors and tenant/occupants demanding or responding? This leads to the importance of LEED certification branding or labelling.

6.1.3. Question 3: Could Buildings without Certification Achieve Same Returns?

All interviewees were asked: 'Could a building built in compliance with LEED, but without LEED certification (label), achieve the same market results as a LEED certified (labelled) building?' Interviewee 1 felt the LEED label would attract more buyers. Interviewee 3 felt that just copying LEED rather than meeting LEED standards for certification would not result in the same thing. However, all of them did feel that a non-LEED certified building could achieve the same returns as an otherwise comparable LEED labelled building. This unanimous opinion seems to have originated from the knowledge that the returns currently generated from LEED buildings stem from a reduction in operating costs. Are the key stakeholder firms shifting their business model toward sustainability or is it still bottom-line finance driven? The answers by the expert interviewees might suggest that they may be tilting toward sustainability; but no, all three still focused on financial gains, albeit due to lower operating costs produced by LEED generated efficiency and not gains in higher rent premiums due to the branding appeal of LEED by those seeking office space. Interviewee 1 made a blunt statement with regard to the LEED label: 'Am I going to get more value for it? No, I don't think so.' This question on the energy efficiency of LEED certified buildings was noted earlier in this paper. A review of research by Amiri et al. [92] on LEED certified buildings likewise noted concerns with regard to the performance of LEED certified building in the areas of energy and atmosphere.

6.1.4. Question 4: Will LEED Certification Attract More Purchasers/Tenants?

LEED certification of office buildings is not without the cost of time and effort. We asked the experts representing stakeholder firms if it was worth it—'will LEED certification help attract more potential purchasers/tenants?' All of the interviewees indicated that LEED certification does attract potential purchasers/tenants. The consensus, however, seems to be that potential tenants are not willing to pay a premium on a gross rent basis for a building that is only advancing the development of green buildings. This is a very interesting position. Interviewee 1 stated, 'frankly most are driven by the bottom line.' As Xie et al.'s [93] study on consumers would suggest that pro-environment behavior plays a key role because the number of LEED certified office buildings in Toronto continues to increase. Tenants/occupants are obviously willing to pay a rental premium for space in these offices. A reading of the interview transcripts strongly suggests that the development of LEED buildings is tenant demand-driven. As earlier studies by Eichholtz et al. [86]

found, tenants/occupants were in part conscious of environmental issues and responsive in their real estate choices.

6.1.5. Question 5: Interest in Retrofitting Existing Buildings to LEED Standards?

Policy makers and scholars agree that in urban core areas such as Toronto it will not be green construction but green retrofitting that will most significantly impact climate change and global warming. However, the interviewees were generally quite positive about LEED certification and agreed on the idea of achieving sustainability in new green office buildings. But the question then is, did this support by key stakeholder firms go deeper to embrace retrofitting of existing buildings? Would they make the decision to 'have their building LEED certified and pay the additional costs associated?' Interestingly only Interviewee 2 stated that the firm currently had LEED applications on a number of their office complexes. Interviewee 1 said LEED status would probably not increase occupancy or translate into higher net rent so the answer is no. Interviewee 3 hedged and said maybe for some properties where tenants wanted it. But for others with strong markets (meaning low vacancy rate), long-term leases, utilities performing in peak range, or with serious utilities issues it wouldn't be considered. A case study of the strengths and weakness of retrofitting an existing building is presented by Sun et al. [94] A building belonging to the University of Hong Kong underwent a significant LEED EBOM Gold project retrofit with the expectation of energy savings of 30% however actual building performance was 16% savings on energy. The interviewees were not eager to move in this direction.

6.2. Discussion of Interview Findings

The three expert interviewee responses are summarised in Table 8. The three are similar in their responses with respect to views on the impact of LEED certification on achievable rent, ability to lease/rent, attitudes of prospective tenants and ESG based investing by institutional investors. Their responses explain and provide an answer to the second hypothesis: Hypothesis (H_2) = Financial drivers will be more influential than non-financial drivers in a key stakeholder's decision whether to pursue LEED certification?

Our interpretation of the interviewee responses to the five questions is that they perceive financial drivers as being very influential in the decision but with several caveats. First, the interviewees mention a growing interest in LEED certification by institutional investors that is also reported in research by Eichholtz et al. [43] and Fuerst et al. [33]. Based on their description of investor pressure to build or retrofit green we viewed this as financial influence at the initiating decision point and not as ESG drivers. Second, the interviewees mention the interest in LEED certified office space by current and perspective tenants. They state that tenants are willing to pay rent premiums for LEED certified office space. We also view this as financial influence on the key stakeholders and not environmental or social. The interviewees' comments suggest that tenant interest in LEED is taking as being less of an influence compared to that of investors. This again is viewed as an indication of the strength of financial influence. The rejection of Hypothesis (H_1) seems to have been anticipated by the interviewees. At least for the present, Toronto's urban core Class A office space, whether LEED or not, has a very low vacancy rate and asking rent is not an issue. Third, interviewees' view of owner landlords—a group that also includes institutional investors and real estate developers—was quite interesting. While acknowledging their influence and importance the interviewees seem to scorn owners as out-front innovators (a quality of entrepreneurs) leading the change to green buildings. A study by Li et al. [95] suggests a framework for addressing such 'capital' barriers that inhibit the promotion of green buildings.

In general, the results from the interviews are in line with previous findings such as the recent research by Oyedokun et al. [34], Cook et al. [91] and Chegut et al. [96]. As markets continue to evolve and tenants and stakeholders place an increasing amount of importance on sustainability, we are likely to see LEED certification become a more significant factor in the determination of office rents. However, in the future LEED will probably become a standard attribute of Class A office buildings before Class B or C. In addition, as more new and retrofitted LEED certified buildings come on the market, future studies can have a larger sample size. This will allow for the investigation of with more subtle impacts on the model and LEED certification as a significant contributor to property value.

7. Conclusions, Limitations and Future Directions

7.1. Conclusions

This study contributes to the literature on green certification and labelling of office buildings in global cities. This study focuses on office buildings in the CBD urban core of Toronto, an area where the financial valuation of green labelling has not attracted sufficient research. The first stage was designed to obtain quantitative data to determine if LEED certification (independent variable) has a significant impact on net asking rent (dependent variable). Pearson correlation and linear regression analysis did not find LEED certification to be statistically significant in explaining the variance in net asking rent (market value). Although this is an exploratory study it is important because it provides an understanding of office buildings in the urban core of Toronto. It is also helpful in setting parameters for future studies on the decisions by key office building stakeholders to build or retrofit green.

The second stage of this study was designed to examine the financial and non-financial factors that influence the decisions by key stakeholders to pursue a green agenda. Qualitative data were obtained from interviews with senior executives who are engaged in Toronto's office building sector. The interviewees provided valuable data to assess if financial drivers are the deciding factor in the decision to pursue LEED certification. They report that institutional investors are most influenced by clients who will only invest in an office building if it meets green standards. Owner landlords are primarily focused on the long-term benefits of LEED certification based on energy efficiency gains that then translate into lower GHG emissions and federal carbon taxes. With respect to real estate developers who are directly involved in the construction of new and existing office buildings, they are responsive to investor preferences, tenant trends and owner concerns. On the issue of whether their decisions are impacted by financial considerations the answer is yes. However, the answer is no on the one factor of LEED certification but on the overall impact of factors/benefits that impact their bottom line. In general, key real estate development stakeholders in Toronto seem unwilling to 'step out and lead' with a focus only on green, but they are also unwilling to be left without green office buildings in their portfolio. Basically, they are cautious in their commitment to support an all-out endorsement of a green agenda. It appears that they are ruled by long-standing and slowly evolving traditions—where the pack stays together, few leap-ahead and few are left behind.

7.2. Limitations

Despite efforts made to ensure the best possible design for this study, there were several areas of limitation. First, Toronto's urban core is one that has the most LEED certified office buildings in Canada. Although this made it an ideal test site because it provided a sufficient sample it also had one limitation. Most of the office buildings with LEED certification are rated as Class A, already command premium rent, and are located in the most prime office area in the city. The sample of conventional (non-LEED) office buildings that met our criteria were primarily Class B and a few Class C buildings located outside of the prime urban core. Although the office buildings included in our sample were all located in the geographic area selected for the study, their rent gradients (CROW FLY) were not. This suggests that a study of the submarkets within Toronto would increase the number of office buildings with a sample of different ratings (Class A, B, and C). This would be similar to the study by Oyedokun et al. [34] that looked at the green office market in the context of local markets. Another study could be designed to focus only on Class A office buildings in the prime urban core of Toronto. However, in such a study the significance of LEED certification could easily be overshadowed by other factors that are associated with leasing office space in the market's prime location.

A second limitation is the fact that importance and growth in numbers of green office buildings in Toronto is still developing. Data and analysis in this study, however, do provide an understanding of the increasingly important role that LEED certification has on the office building sector. This will continue in the future. As this movement matures, longitudinal studies related to changes in costs and benefits in LEED certified construction, measured over a period of time and over varying economic conditions, would provide valuable assistance for addressing the environmental issues of climate change.

Third, there were limitations in terms of how much specific information we could ask interviewees to provide. The three Toronto real estate development executives volunteered as anonymous industry experts, not as informants representing their firms. This restricted depth of questioning on specific examples or on their own positions and firms. A survey questionnaire of top tier executives from the dozen or so key firms in the Toronto office building market might have produced data on drivers and influencers for statistical analysis but obtaining that information is very difficult. This was considered as the research design was developed but the reality of actually getting a large enough sample of office building developers seemed unachievable. It also seemed that personal interviews with a few of the executives in different firms might only produce 'sound good statements' rather than actual reality [43,45].

Finally, it would have been informative to include actual rent paid per square foot. As stated earlier, most of the firms view information related to tenant lease contracts as proprietary and confidential. The best alternative was to use average net asking rental rates for each building as a proxy. Future studies could try to include information related to actual transacted leases to construct a model that explains the variance between actual transacted rents and hedonic independent variables such as LEED certification.

7.3. Future Directions

Although this is the first study that looked at issues related to LEED certified buildings in Toronto, we hope that scholars will add to our research in ways suggested in the limitation section. These efforts are needed, especially as the challenges of climate change and global warming increase the pressure to build and retrofit green. Interest and pressure are mounting on key stakeholders to address the broader accompanying aspects of sustainability such as social and corporate governance factors that are gaining traction in the public and political arena. Considering these developments, we feel our choice of that the expert interview format to gain first-hand information is a promising direction for future research.

The framework of this study—looking at the issue from a context dimension (place-based gradient setting) and process dimension (real estate development process)—was an excellent approach for understanding the market and decision dynamics on the issue of building and retrofitting green. Perhaps the question for further study is, should these issues be viewed at the individual executive level (motivation and ethics) or firm level (value proposition and business model)? Either approach would provide a theoretical platform to understand the drivers and barriers that impact this decision. Further research could look at the cascading decisions and stakeholders involved in the implementation of this decision.

Author Contributions: The authors contributed equally to the manuscript preparation: writing, revising and editing. All authors have read and agreed to the published version of the manuscript.

Funding: This research was funded by the Social Sciences and Humanities Research Council of Canada.

Institutional Review Board Statement: University Research Grants and Ethics Services provided approval through the Social Sciences and Humanities Research Ethics Board, which operates in accordance with the Tri-Council Policy Statement and other applicable laws and regulations in Ontario.

Informed Consent Statement: Informed consent was obtained from the three interviewees involved in the study.

Data Availability Statement: Interview data are provided in the Appendix A, statistically analysed data may be obtained by contacting the sources identified.

Conflicts of Interest: The authors declare no conflict of interest.

Appendix A. Transcription of Expert Interviewees

Interviewee 1

Profile: serves as Vice President in one of Canada's largest landlord developer firms. In that role is responsible for the origination and execution of office, industrial, and land transactions, together with major property portfolios across Canada.

In your experience, do you feel that LEED certification has a significant impact on achievable net rents?

Having a LEED building certainly does cut down the operating expenses, allowing you to achieve a greater net rent while still remaining competitive with other buildings on a gross basis. You might be higher on a net basis, but you will be lower on a gross basis because you are able to achieve some net savings that you can pass on to the tenant.

But above and beyond that—take 18 York Street for example—that is a building that Gowling WLG is building on behalf of BCIMC [British Columbia Investment Management Corporation], and their big tenant they have right now is PWC. When these guys pulled the trigger back in 2008 at a point when the market had already shifted down into tenant favour. Nobody was leasing space. There was a credit crunch. Investment transactions were few and far between and yet BCIMC still went ahead to develop that building. And the reason why was because they wanted to be able to say that they owned a downtown building that was LEED certified. Their thought was that large tenants on a go forward basis—banks, insurance firms, law firms—will make it a requirement. That any space that they lease has to be in a LEED building that is reflective of their overall objective with respect to the environment in their business plan model. So yeah, I think that over time you are going to see that rents are only going to go in one direction for LEED buildings.

Do you think that LEED buildings are going to be considered the new A class?

Tenants are only concerned with the gross rents. They do not care what the net rent is, they do not really care what the taxes and operating costs are. They want to know what their gross rents are. So as long as their gross rents are competitive to Class A buildings and they know that their carbon footprint is less and they are doing a good thing for the environment they'll go there. Although some tenants will make that a requirement and will pay more just to be in that building. But, if you are able to track a comparison between LEED buildings and comparable Class A buildings, I think you are going to see at least a $2–3/sq. ft. savings, and I think that you will be able to make up for that in net rent.

Above and beyond additional rent savings, do you think that from a marketing perspective, over the long term, LEED buildings will be able to achieve significantly higher rents?

No. There is an inherent value that you are adding. From a liquidity standpoint, let's say, if I am the broker and I have got two buildings I can sell. One is the triple A office building downtown, and the other is similar building that is LEED certified I believe the LEED building will attract more buyers because people are feeling conscientious. Am I going to get more value for it? No, I don't think so.

Do you think that from a sales standpoint LEED certification is going to help you attract more potential purchasers?

Yes. I believe that, but I do not think that will necessarily translate into more value. And that relates back to your earlier question where you asked me if I thought a tenant would be willing to pay more for a LEED building, and I said no. And the reason I say no is that tenants have their own going concerns that they are worried about and rent represents a really big part of their obligations, their costs, especially if you are renting four floors of downtown office space and frankly they are driven by the bottom line. Yeah, you do have some big companies such as maybe PWC that were motivated by the fact that it was a LEED building, but I think it was just that they were the lead tenant in that project, so undoubtedly GWL gave them a bit of a concession to get them in and then they could attract others.

If I had a Class B building that I was planning on retrofitting to bring it up to date with the aim of reducing my operating costs in an effort to increase my achievable net rent, do you think I should undertake to have the building LEED certified and pay the additional costs associated with that? Or do you say if you can reduce additional costs for your prospective tenants then that is good enough?

I would say—and it's unfortunate—it's probably the latter. We are selling a building right now where they have BOMA BEST certification, where they have acknowledged that they have been able to reduce operating costs through efficiencies. It is a Class A building and it is in the suburbs, but at the end of the day if they were able to attain LEED status I don't think that would increase their occupancy, or necessarily translate into more net rents, and it's unfortunate, but I really don't think that is the case.

Interviewee 2

Profile: serves as Senior Vice President for one of Canada's largest landlord developers. In that role is involved in informing the decision-making that provides the company with strategic direction.

I understand you have a building in downtown Toronto that you are getting LEED certified, why did you decide to prescribe to LEED certification, and why this building specifically?

It goes back some years, while we had the site under contract and we closed in December of that year. Imagine a board meeting where I was making a presentation to three owners in September. Three different owners: The first one was the Menkes Family, who had a 20 percent interest in the property. Their comment was—we have no clue what LEED is, and this was before LEED was even a thing in Canada—we understand in terms of the world sustainability is going to be necessary. We are responsible developers, so if you think it is the right thing to do then it will be one of our specifications and we will figure it out as we go. That's 20 percent. Then the hospitals of Ontario Pension Plan, who owned 50 percent. They say—we also don't know what it is because we haven't seen anything, no building exists in Canada. But we, representing our pensioners, all 40,000 of them and wish to respect sustainability in everything we do. That's 50 percent. The Harvard endowment fund, who represented 20 percent, said—if this is not a LEED building we will not invest, simply because we are setting the standards in terms of regulations of the sort worldwide. Our people are, our graduates are, and sustainability is a must. That is where it all started. So, we started as responsible citizens, and then as investors, and then as owners of real estate in Canada. The LEED manuals didn't even hit the table until two years later. So, we were ahead of the curve, and in hindsight every single tenant who walked into the building asked about our sustainability platform. Every single tenant who walked into the building wanted a green platform.

Seeing the success, you have achieved with this building, are their plans for the certification of others?

We have another office development downtown and the building will be somewhere in the 900,000 square feet range and we too will be LEED Gold core and shell. Moving forward it is simply the only standard we have.

Are you only looking to certify new developments, or are you looking at retrofits as well?

In terms of existing buildings, we have numerous applications in for LEED on a number of our office complexes. No industrial.

Is there a reason why you are not for industrial but are for office buildings?

The tenants are more interested in it. And on industrial it is next to impossible to achieve on an existing industrial space. You can achieve it moving forward with a new LEED development, but moving backwards, no.

Comparing your LEED certified building to others in your portfolio and others you compete with in that market, do you think that the LEED certification has been a factor in achievable gross rent.

Zero. Now let's qualify that. There is no history. Everything we put into 25 York was a projection. People had to believe it was going to happen. I had no ability to get any higher rent than another office tower in downtown Toronto which doesn't have it, because there was no proof. Many a time tenants ask—Well, how do you know? and we would reply—Because our engineers tell us. And they reply—SO, you know the Prime Minister of Canada tells us we are going to be debt free too... show me! Ask that question in 5–10 years. Then the green buildings will stand out.

So, do you think that moving forward LEED is going to be the new standard?

Definitely! As markets go forward and, from an appraisal point of view, there will be a factor built into all appraisals which give green buildings more value because they will last longer.

What do you think the impact of LEED certification will be as markets fluctuate?

If it is a tenant's market, LEED will win first. If it is a landlord's market, nobody gives a shit.

Additional discussion.

Every single tenant that moved into our LEED building insisted on construction to LEED standard. When asked: are you going to certify? One tenant said that, according to their corporate platform, they wished to be one of the top 50 sustainable companies in the world. So, they went through a certification process. Every other tenant said—I don't need a certification; I don't need to pay for all that bullshit. I get nothing for the certificate on the wall. I just want to know myself I did the right thing.

Apart from the reduction in operating costs, do you find any value in the LEED label itself?

Not today.

So, if you had an otherwise comparable property, are you saying that you would not bother getting the building LEED certified?

You asked me do I perceive any value! Me the landlord. Examining it from the tenant's point of view. Not one tenant believes one landlord! Therefore, I have to get a certification by a third party in order for the tenant to believe me. The tenants perceive value in the buildings. Long term, LEED count. It may not be called LEED in 10 or 15 years, but sustainability counts. And what is happening with these buildings is that, over and over, the manufacturers of materials and equipment are pursuing certification themselves because they are listening to the politicians saying that it is coming, we have to save the world. More importantly they are listening to their kids.

On top of that, every single pension plan, the Caisse (Caisse de depot et placement du Québec)amaz, [unintelligible], Teachers Pension Plan of Ontario, BCIMC, IMCO [Investment Management Corporation of Ontario]. They have said, they have decreed, they have policies, if we are developing new office buildings, they shall be green. Representing their pensioners, their pensioners are average everyday people. It started with BCIMC out of British Columbia, as it should, that is where the green movement started.

Interviewee 3

Profile: serves as the National Director of Sustainability and Energy Management for one of Canada's largest landlord developer firms. In that role informs company decision-making related to making the company more environmentally responsible and energy efficient.

Do you have any LEED certified buildings?

Yes, we do.
Are they new builds or retrofits?
New builds, but we are also going to have a couple of retrofits coming up this year.

With respect to the buildings that you already have certified and the ones that are soon to be certified, why did you decide to certify these buildings?

Two reasons. At the time LEED was beginning to be considered a competitive advantage, and for the buildings we decided to have certified it was not a complicated process and the cost premium was not that significant.

When you talk about competitive advantage, are you talking about making it easier to lease the space thereby reducing vacancy, or are you talking about an ability to achieve higher gross rents.

When we develop the building pro forma, we put together the project and we start marketing it so that the building is already tenanted by the time it is delivered. So, as we were marketing the property some of our clients liked the fact that it was LEED certified. They see it through two perspectives: One is the PR. Marketing wise they see it as something they can make use of and portray their companies as more environmentally conscious with a higher CSR status.

The second is that there is a higher level of predictability when it comes to utilities. So, they know that if they are coming into a certified building there are higher environmental and ethical standards, and that the building is going to perform closer to what they have budgeted for in future years. So, the level of certainty is important, especially for those companies where real estate is a big component of their finances.

Did LEED certification meet your expectations in terms of competitive advantage for your buildings?

Yes. Additionally, some of the tenants in our non-LEED certified buildings are looking for LEED certified space.

Are you noticing LEED as a factor for other potential tenants?

Yes. When we are building a property, potential tenants will ask—do you guys plan on building it to LEED certified standards; and we say—yes, and that becomes part of the negotiation. And the big thing is that we wouldn't do it any other way. We wouldn't do it any other way because it is now the market expectation to have your building LEED certified.

Are you saying that from now on when you build it is going to be LEED certified?

Yes, pretty much. When we had decided to certify one of our buildings about 10, maybe almost 15 years ago, it was a marketing advantage, while now it is becoming the standard and I am not sure how far down the road it is going to be detrimental in some cases where we will have to LEED certify.

Do you think that you could have built to LEED certification standards without getting the buildings certified and achieved the same results?

Yes, however collecting the data and following the protocols is not as efficient if you do not have to comply with a certification. The moment you say that we need to do this to get the certification things start to happen faster. Even if you don't have the certification you should be able to do the work to that standard, but at the end it doesn't seem to work that way. The certification process acts as a guide throughout the development process.

Do you plan on retrofitting all your buildings as well to be LEED certified?

Not all of them. There are some buildings where it makes a lot of sense, where we have tenants that want to be in a certified building, some buildings where we have challenges with utilities, and so on. But there are properties where there is no need for it. It could be because they are in a very strong market, because we have long-term leases, because the building is performing in peak range, so in those cases it wouldn't be necessary to pursue certification.

Are there some buildings where you will not look to retrofit because of the difficulty in getting the building to a state where it complies with LEED standards?

Yes. The capital required to bring some buildings up to LEED standards makes it impossible for us to get them certified. There are also issues with retrofitting existing industrial and multi-residential properties, so currently we are only retrofitting office buildings. However, we are pursuing LEED certification for all new builds.

References

1. United Nations. Climate Change Annual Report 2019. Available online: https://unfccc.int/sites/default/files/resource/unfccc_annual_report_2019.pdf (accessed on 21 December 2020).
2. Rana, I.A.; Routray, J.K.; Younas, Z.I. Spatiotemporal dynamics of development inequalities in Lahore city region, Pakistan. *Cities* **2020**, *96*, 102418. [CrossRef]
3. Carlson, K.; Pressnail, K.D. Value impacts of energy efficiency retrofits on commercial office buildings in Toronto, Canada. *Energy Build.* **2018**, *162*, 154–162. [CrossRef]
4. Wang, Z.; Li, H.; Zhang, B.; Tian, X.; Zhao, H.; Bai, Z. The greater the investment, the greater the loss?—Resource traps in building energy efficiency retrofit (BEER) market. *Res. Conserv. Recycl.* **2021**, *168*, 105459. [CrossRef]
5. Calvet, S. Morguard Revitalizes 60 Bloor West. *Urban Toronto*, 4 March 2021. Available online: https://urbantoronto.ca/news/2021/03/morguard-revitalizes-60-bloor-west (accessed on 18 March 2021).
6. Government of Canada. 2019 Greenhouse Gas Emissions. Available online: https://www.canada.ca/en/environment-climate-change/services/environmental-indicators/greenhouse-gas-emissions.html (accessed on 18 March 2021).
7. City of Toronto. 2018 Greenhouse Gas Emissions Inventory. Available online: https://www.toronto.ca/wp-content/uploads/2020/12/9525-2018-GHG-Inventory-Report-Final-Published.pdf (accessed on 18 March 2021).
8. Lewington, J.; Toronto office tower takes the road to zero carbon. The Globe and Mail, 19 January 2021. Available online: https://www.theglobeandmail.com/business/industry-news/property-report/article-toronto-office-tower-takes-the-road-to-zero-carbon/ (accessed on 18 March 2021).
9. Wen, B.; Musa, S.N.; Onn, C.C.; Ramesh, S.; Liang, L.; Wang, W.; Ma, K. The role and contribution of green buildings on sustainable development goals. *Build. Environ.* **2020**, *185*, 107091. [CrossRef]
10. Lucuik, M.; Trusty, W.; Larsson, N.; Charette, R. *A Business Case for Green Buildings in Canada*; Morrison Hershfield: Ottawa, ON, Canada, 2005.
11. US Green Building Council. Available online: https://www.usgbc.org/articles/us-green-building-council-announces-top-10-countries-and-regions-leed-green-building (accessed on 21 December 2020).
12. Addae-Dapaah, K.; Hiang, L.K.; Sharon, N.Y.S. Sustainability of sustainable real property development. *J. Sustain. Real Estate* **2009**, *1*, 203–225. [CrossRef]
13. Mangialardo, A.; Micelli, E.; Saccani, F. Does sustainability affect real estate market values? Empirical evidence from the office building in Milan (Italy). *Sustainability* **2018**, *11*, 12. [CrossRef]
14. Mok, K.Y.; Shen, G.Q.; Yang, R. Stakeholder complexity in large scale green building projects. *Eng. Const. Arch. Manag.* **2018**, *25*, 1454–1474. [CrossRef]
15. Wadu Mesthrige, J.; Chan, H.T. Environmental certification schemes and property values: Evidence from the Hong Kong prime commercial office market. *Int. J. Strat. Prop. Manag.* **2019**, *23*, 81–95. [CrossRef]
16. Simons, R.A.; Feltman, D.C.; Malkin, A.A. When would driverless vehicles make downtown parking unsustainable, and where would the driverless car fleet rest during the day? *J. Sustain. Real Estate* **2018**, *10*, 3–32. [CrossRef]
17. Graaskamp, J.A. Fundamentals of real estate development. *J. Prop. Val. Invest.* **1992**, *10*, 619–639. [CrossRef]
18. Goubran, S.; Cucuzzella, C. Integrating the sustainable development goals in building projects. *J. Sustain. Res.* **2019**, *1*, 190010.
19. Bazhenova, E.; Weske, M. Deriving decision models from process models by enhanced decision mining. In *International Conference on Business Process Management*; Springer: Cham, Switzerland, 2016; pp. 444–457. Available online: https://www.researchgate.net/profile/Ekaterina-Bazhenova-2/publication/305520086_Deriving_Decision_Models_from_Process_Models_by_Enhanced_Decision_Mining/links/5afd9d780f7e9b98e09ede02/Deriving-Decision-Models-from-Process-Models-by-Enhanced-Decision-Mining.pdf (accessed on 22 March 2021).
20. Timmons, J.A.; Spinelli, S.; Ensign, P.C. *New Venture Creation*, Canadian ed.; McGraw-Hill: Toronto, ON, Canada, 2010.
21. Teece, D.J. Business models, business strategy and innovation. *Long Range Plan.* **2010**, *43*, 172–194. [CrossRef]
22. Geissdoerfer, M.; Vladimirova, D.; Evans, S. Sustainable business model innovation: A review. *J. Clean. Prod.* **2018**, *198*, 401–416. [CrossRef]
23. Preghenella, N.; Battistella, C. Exploring business models for sustainability: A bibliographic investigation of the literature and future research directions. *Bus. Strat. Environ.* **2021**. [CrossRef]

24. Osterwalder, A.; Pigneur, Y. Business Model Canvas. Self-Published. 2010. Available online: http://www.cenda.cl/images/descargas/sanvicente.pdf (accessed on 31 October 2020).
25. Joyce, A.; Paquin, R.L. The triple layered business model canvas: A tool to design more sustainable business models. *J. Clean. Prod.* **2016**, *135*, 1474–1486. [CrossRef]
26. Comin, L.C.; Aguiar, C.C.; Sehnem, S.; Yusliza, M.Y.; Cazella, C.F.; Julkovski, D.J. Sustainable business models: A literature review. *Benchmarking Int. J.* **2019**, *27*, 2028–2047. [CrossRef]
27. Zaccack, N.R. How Real Estate Developers Define and Implement Their Social Impact Goals through the Real Estate Development Process. Ph.D. Thesis, Master of Science in Real Estate Development, Massachusetts Institute of Technology, Cambridge, MA, USA, 2020.
28. Brigham, E.F. The determinants of residential land values. *Land Econ.* **1965**, *41*, 325–334. [CrossRef]
29. Eichholtz, P.; Kok, N.; Quigley, J. The economics of green building. *Rev. Econ. Stats.* **2013**, *95*, 50–63. [CrossRef]
30. Devine, A.; Kok, N. Green certification and building performance: Implications for tangibles and intangibles. *J. Portf. Manag.* **2015**, *41*, 151–163.
31. Deng, Y.; Wu, J. Economic returns to residential green building investment: The developers' perspective. *Region. Sci. Urb. Econ.* **2013**, *47*, 35–44. [CrossRef]
32. Chegut, A.; Eichholtz, P.; Kok, N. Supply, demand and the value of green buildings. *Urban Stud.* **2014**, *51*, 22–43. [CrossRef]
33. Fuerst, F.; Gabrieli, T.; McAllister, P. A green winner's curse? Investor behavior in the market for eco-certified office buildings. *Econ. Model.* **2017**, *61*, 137–146. [CrossRef]
34. Oyedokun, T.; Jones, C.; Dunse, N. The growth of the green office market in the UK. *J. Euro. Real Estate Res.* **2015**, *8*, 267–284. [CrossRef]
35. Falkenbach, H.; Lindholm, A.L.; Schleich, H. Review articles: Environmental sustainability: Drivers for the real estate investor. *J. Real Estate Lit.* **2010**, *18*, 201–223. [CrossRef]
36. Darko, A.; Zhang, C.; Chan, A.P. Drivers for green building: A review of empirical studies. *Habitat Int.* **2017**, *60*, 34–49. [CrossRef]
37. Zhang, X. Green real estate development in China: State of art and prospect agenda—A review. *Ren. Sustain. Energy Rev.* **2015**, *47*, 1–13. [CrossRef]
38. Whitney, S.; Dreyer, B.C.; Riemer, M. Motivations, barriers and leverage points: Exploring pathways for energy consumption reduction in Canadian commercial office buildings. *Energy Res. Soc. Sci.* **2020**, *70*, 101687. [CrossRef]
39. Leskinen, N.; Vimpari, J.; Junnila, S. A review of the impact of green building certification on the cash flows and values of commercial properties. *Sustainability* **2020**, *12*, 2729. [CrossRef]
40. Liang, X.; Peng, Y.; Shen, G.Q. A game theory-based analysis of decision making for green retrofit under different occupancy types. *J. Clean Prod.* **2016**, *137*, 1300–1312. [CrossRef]
41. Cohen, C.; Pearlmutter, D.; Schwartz, M. A game theory-based assessment of the implementation of green building in Israel. *Build. Environ.* **2017**, *125*, 122–128. [CrossRef]
42. Fan, K.; Hui, E.C. Evolutionary game theory analysis for understanding the decision-making mechanisms of governments and developers on green building incentives. *Build. Environ.* **2020**, *179*, 106972. [CrossRef]
43. Eichholtz, P.; Kok, N.; Quigley, J.M. Doing well by doing good? Green office buildings. *Amer. Econ. Rev.* **2010**, *100*, 2492–2509. [CrossRef]
44. Fennis, B.M.; Adriaanse, M.A.; Stroebe, W.; Pol, B. Bridging the intention–behavior gap: Inducing implementation intentions through persuasive appeals. *J. Consum. Psych.* **2011**, *21*, 302–311. [CrossRef]
45. Zhang, G.; Zhang, Y.; Tian, W.; Li, H.; Guo, P.; Ye, F. Bridging the intention–behavior gap: Effect of altruistic motives on developers' action towards green redevelopment of industrial brownfields. *Sustainability* **2021**, *13*, 977. [CrossRef]
46. Canada Green Build Council LEED Canada Rating System. Available online: https://www.cagbc.org/CAGBC/Programs/LEED/LEED_Canada_Rating_System/LEED_Canada_Rating_System.aspx (accessed on 11 November 2020).
47. Eco Home. Available online: https://www.ecohome.net/guides/1217/canada-ranks-as-the-top-country-for-leed-green-building-for-second-year-in-a-row/ (accessed on 11 November 2020).
48. Duggan, E.; Toronto's 5 Largest New or Under Construction Office Buildings: With Toronto's Office Vacancy Rate at About One Percent, Developers Have about 9.4 Million Square Feet of New Office Projects Under Way. Vancouver Sun, 12 June 2019. Available online: https://vancouversun.com/business/commercial-real-estate/torontos-5-largest-new-or-under-construction-office-buildings (accessed on 18 March 2021).
49. McClintock, A. Did the Pandemic Crush Commercial Real Estate? We Asked a Market Expert. *Toronto Life* . 24 June 2020. Available online: https://torontolife.com/real-estate/did-the-pandemic-crush-commercial-real-estate-we-asked-a-market-expert/ (accessed on 18 March 2021).
50. Statista. Available online: https://www.statista.com/statistics/802500/asking-rent-per-sf-office-space-toronto-by-submarket/ (accessed on 18 March 2021).
51. Kibert, C.J. *Sustainable Construction: Green Building Design and Delivery*, 4th ed.; John Wiley & Sons: Hoboken, NJ, USA, 2016.
52. Sayce, S.; Ellison, L.; Parnell, P. Understanding investment drivers for UK sustainable property. *Build. Res. Info.* **2007**, *35*, 629–643. [CrossRef]
53. Lippiatt, B.C. Selecting cost-effective green building products: BEES approach. *J. Constr. Eng. Manag.* **1999**, *125*, 448–455. [CrossRef]

54. Yasinta, R.B.; Utomo, C.; Rahmawati, Y. A literature review of methods in research on green building cost analysis. In *IOP Conference Series: Materials Science and Engineering*; IOP Publishing: Bristol, UK, 2020; Volume 930, p. 012014. Available online: https://iopscience.iop.org/article/10.1088/1757-899X/930/1/012014/pdf (accessed on 21 March 2021).
55. Bartlett, E.; Howard, N. Informing the decision makers on the cost and value of green building. *Build. Res. Inf.* **2000**, *28*, 315–324. [CrossRef]
56. Chegut, A.; Eichholtz, P.; Kok, N.; Quigley, J.M. The value of green buildings: New evidence from the United Kingdom. *ERES 2010 Proc.* Available online: https://immobilierdurable.eu/images/2128_uploads/Chegut_Eichholtz_Kok_green_value_in_the_uk.pdf (accessed on 20 March 2021).
57. Zhang, L.; Wu, J.; Liu, H. Turning green into gold: A review on the economics of green buildings. *J. Clean Prod.* **2018**, *172*, 2234–2245. [CrossRef]
58. Hsieh, H.C.; Claresta, V.; Bui, T.M.N. Green building, cost of equity capital and corporate governance: Evidence from US real estate investment trusts. *Sustainability* **2020**, *12*, 3680. [CrossRef]
59. Hwang, B.G.; Tan, J.S. Green building project management: Obstacles and solutions for sustainable development. *Sustain. Dev.* **2012**, *20*, 335–349. [CrossRef]
60. Tsalis, T.A.; Malamateniou, K.E.; Koulouriotis, D.; Nikolaou, I.E. New challenges for corporate sustainability reporting: United Nations' 2030 Agenda for sustainable development and the sustainable development goals. *Corp. Soc. Resp. Environ. Manag.* **2020**, *27*, 1617–1629. [CrossRef]
61. Pivo, G. Responsible property investing: What the leaders are doing. *J. Prop. Invest. Fin.* **2008**, *26*, 562–576. [CrossRef]
62. Oyedokun, T.B. Green premium as a driver of green-labeled commercial buildings in the developing countries: Lessons from the UK and US. *Int. J. Sustain. Built. Environ.* **2017**, *6*, 723–733. [CrossRef]
63. Samad, N.S.A.; Abdul-Rahim, A.S.; Yusof, M.J.M.; Tanaka, K. Impact of green building certificate on firm's financial performance. *IOP Conf. Ser. Earth Environ. Sci.* **2020**, *549*, 012076. Available online: https://iopscience.iop.org/article/10.1088/1755-1315/549/1/012076/pdf (accessed on 21 March 2021). [CrossRef]
64. Qiu, Y.; Su, X.; Wang, Y.D. Factors influencing commercial buildings to obtain green certificates. *Appl. Econ.* **2017**, *49*, 1937–1949. [CrossRef]
65. Kats, G.; Alevantis, L.; Berman, A.; Mills, E.; Perlman, J. The Costs and Financial Benefits of Green Buildings: A Report of California's Sustainable Building Task Force, 134 (October, 2003). Available online: https://noharm-uscanada.org/sites/default/files/documents-files/34/Building_Green_Costs_Benefits.pdf (accessed on 11 May 2021).
66. Lützkendorf, T.; Lorenz, D. Sustainable property investment: Valuing sustainable buildings through property performance assessment. *Build. Res. Info.* **2005**, *33*, 212–234. [CrossRef]
67. Santos, R.; Costa, A.A.; Silvestre, J.D.; Pyl, L. Integration of LCA and LCC analysis within a BIM-based environment. *Autom. Construct.* **2019**, *103*, 127–149. [CrossRef]
68. Warren-Myers, G.; Bienert, S.; Warren, C. Valuation and sustainability are rating tools enough? In Proceedings of the European Real Estate Society Conference, Stockholm, Sweden, 24–27 June 2009.
69. Millington, A.F. *An Introduction to Property Valuation*, 5th ed.; Routledge Taylor & Francis Group: London, UK, 2013.
70. Seyler, N.J. *Sustainability and the Occupant: The Effects of Mindfulness and Environmental Attitudes on Real Estate User Behaviors*; Springer Fachmedien: Wiesbaden, Germany, 2020.
71. Singh, A.; Syal, M.; Grady, S.C.; Korkmaz, S. Effects of green buildings on employee health and productivity. *Am. J. Pub. Health* **2010**, *100*, 1665–1668. [CrossRef]
72. Esfandiari, M.; Zaid, S.M.; Ismail, M.A.; Aflaki, A. Influence of indoor environmental quality on work productivity in green office buildings: A review. *Chem. Eng. Trans.* **2017**, *56*, 385–390.
73. Das, P.; Wiley, J.A. Determinants of premia for energy-efficient design in the office market. *J. Prop. Res.* **2014**, *31*, 64–86. [CrossRef]
74. Fuerst, F.; van de Wetering, J. How does environmental efficiency impact on the rents of commercial offices in the UK? *J. Prop. Res.* **2015**, *32*, 193–216. [CrossRef]
75. Oyedokun, T.B.; Dunse, N.; Jones, C. The impact of green premium on the development of green-labeled offices in the UK. *J. Sustain. Real Estate* **2018**, *10*, 81–108. [CrossRef]
76. Clapp, J.M. The intrametropolitan location of office activities. *J. Reg. Sci.* **1980**, *20*, 387–399. [CrossRef]
77. Hough, D.; Kratz, C. Can 'good' architecture meet the market test? *J. Urban Econ.* **1983**, *14*, 40–54. [CrossRef]
78. Cannaday, R.E.; Kang, H.B. Estimation of market rent for office spaces. *Real Estate Appraiser Anal.* **1984**, *50*, 67–72.
79. Brennan, T.P.; Cannaday, R.E.; Colwell, P.F. Office rent in the Chicago CBD. *Real Estate Econ.* **1984**, *12*, 243–260. [CrossRef]
80. Glascock, J.L.; Jahanian, S.; Sirmans, C.F. An analysis of office market rents: Some empirical evidence. *Amer. Real Estate Urb. Econ. Assoc. J.* **1990**, *18*, 105–119. [CrossRef]
81. Mills, E.S. Office rent determinants in the Chicago area. *Real Estate Econ.* **1992**, *20*, 273–287. [CrossRef]
82. Dunse, N.; Jones, C. A hedonic price model of office rents. *J. Prop. Valuat. Invest.* **1998**, *16*, 297–312. [CrossRef]
83. CBRE and Maastricht University. International Green Building Adoption Index 2018. Available online: https://static1.squarespace.com/static/57fa8eb0b3db2b529d44fa66/t/5ade1f3188251b9506c7b346/1524506429700/IGBAI2018.pdf (accessed on 18 March 2021).
84. Devencore Research; Toronto downtown office market, demand for top-tier space drives downtown Toronto office market. Personal Communication, 2017.

85. Building Owners and Managers Association International. Building Class Definitions. Available online: https://www.boma.org/BOMA/Research-Resources/Industry_Resources/BuildingClassDefinitions.aspx (accessed on 11 May 2021).
86. Eichholtz, P.; Kok, N.; Quigley, J.M. Why do companies rent green? Real property and corporate social responsibility. In *Real Property and Corporate Social Responsibility (August 20, 2009). Program on Housing and Urban Policy Working Paper*; University of California; Berkeley (W09-004). Available online: https://escholarship.org/content/qt7br1062q/qt7br1062q.pdf (accessed on 11 May 2021).
87. Wiley, J.A.; Benefield, J.D.; Johnson, K.H. Green design and the market for commercial office space. *J. Real Estate Financ. Econ.* **2010**, *41*, 228–243. [CrossRef]
88. Fuerst, F.; McAllister, P. Green noise or green value? Measuring the effects of environmental certification on office values. *Real Estate Econ.* **2011**, *39*, 45–69. [CrossRef]
89. Bera, A.K.; Uyar, S.G.K. Local and global determinants of office rents in Istanbul. *J. Eur. Real Estate Res.* **2019**, *12*, 227–249. [CrossRef]
90. Döringer, S. 'The problem-centered expert interview'. Combining qualitative interviewing approaches for investigating implicit expert knowledge. *Int. J. Soc. Res. Methodol.* **2021**, *24*, 265–278. [CrossRef]
91. Cook, J.; Govett, C.; Murray, M. Supreme green: The effect of LEED certification levels on net and gross lease rental rates in commercial offices. University of the South: Sewanee, TN, USA, 2020. Available online: https://dspace.sewanee.edu/bitstream/handle/11005/21695/GovettSupremeSS2020.pdf?sequence=1&isAllowed=y (accessed on 20 June 2021).
92. Amiri, A.; Ottelin, J.; Sorvari, J. Are LEED-certified buildings energy-efficient in practice? *Sustainability* **2019**, *11*, 1672. [CrossRef]
93. Xie, X.; Lu, Y.; Gou, Z. Green building pro-environment behaviors: Are green users also green buyers? *Sustainability* **2017**, *9*, 1703. [CrossRef]
94. Sun, X.; Gou, Z.; Lu, Y.; Tao, Y. Strengths and weaknesses of existing building green retrofits: Case study of a LEED EBOM gold project. *Energies* **2018**, *11*, 1936. [CrossRef]
95. Li, Q.; Long, R.; Chen, H.; Chen, F.; Wang, J. Visualized analysis of global green buildings: Development, barriers and future directions. *J. Clean. Prod.* **2020**, *245*, 118775. [CrossRef]
96. Chegut, A.; Eichholtz, P.; Kok, N. The price of innovation: An analysis of the marginal cost of green buildings. *J. Environ. Econ. Manag.* **2019**, *98*, 102248. [CrossRef]

MDPI

Article

Materiality Matrix Use in Aligning and Determining a Firm's Sustainable Business Model Archetype and Triple Bottom Line Impact on Stakeholders

Valeska V. Geldres-Weiss [1,*], Nicolás Gambetta [2], Nathaniel P. Massa [3] and Skania L. Geldres-Weiss [4]

1 Department of Management and Economics, Faculty of Law and Business, Universidad de La Frontera, Temuco 4811230, Chile
2 Facultad de Administración y Ciencias Sociales, Universidad ORT Uruguay, Montevideo 11300, Uruguay; gambetta@ort.edu.uy
3 Department of Management, Faculty of Economics, Management and Accountancy, University of Malta, Msida MSD 2080, Malta; nathaniel.massa@um.edu.mt
4 Department of Economics and Business, Facultad de Ciencias Sociales, Empresariales y Jurídicas, Universidad de La Serena, La Serena 1700000, Chile; skania.geldres@userena.cl
* Correspondence: valeska.geldres@ufrontera.cl; Tel.: +56-9-9735-0999

Abstract: The materiality matrix is a tool that helps companies understand how the stakeholders' view of material issues in environmental, social, and economic/governance dimensions influences their value creation process, and creates triple bottom line impacts through shaping their strategic business model elements. Building on the multidimensional definition of materiality, we propose to use the materiality matrix as a tool to aid the transformation of a company's existing traditional business model into a more sustainable one (inside-out approach), and to enable the identification of the most appropriate business model archetype to incorporate innovation into its sustainable business model (outside-in approach). This paper presents the materiality matrix as a new tool to enhance and transpose a company's business model towards sustainability—as illustrated through the analysis of the Viña Concha y Toro business model case. This new tool contributes to sustainable business model literature and stakeholder theory by incorporating the materiality matrix as a gateway to business model innovation, and as a tool to explain the dynamics in the sustainable value creation process and concomitant impact on stakeholders.

Keywords: sustainable business model canvas; sustainable business model archetype; materiality matrix; winery; agri-food sector; sustainability

Citation: Geldres-Weiss, V.V.; Gambetta, N.; Massa, N.P.; Geldres-Weiss, S.L. Materiality Matrix Use in Aligning and Determining a Firm's Sustainable Business Model Archetype and Triple Bottom Line Impact on Stakeholders. *Sustainability* **2021**, *13*, 1065. https://doi.org/10.3390/su13031065

Received: 17 December 2020
Accepted: 15 January 2021
Published: 20 January 2021

Publisher's Note: MDPI stays neutral with regard to jurisdictional claims in published maps and institutional affiliations.

1. Introduction

Five years since its globe-spanning adoption by all 193 United Nations (UN) member states, and a mere decade to its target date—the 2030 Agenda for Development resolution presents an urgent clarion call. This juncture of aroused awareness and incessant demands for sustainability and corporate responsibility, now more than ever, present the perfect opportunity for companies to review their business models in order to understand their value creation processes and gauge whether they are maximising total value to stakeholders—beyond financial imperatives. Financial crises and social calamities (e.g., COVID-19), as well as extreme weather conditions, pose an urgent need for companies to do things differently and responsibly, and to embrace a long-term view of prosperity. To achieve this goal, companies need to develop more holistically sustainable business models. Without changing current business models—in which growth is predicated on selling more goods to more people—environmental stresses will increase business risks and costs—mitigating and ultimately compromising essential fundamentals of sustainability. Some companies are examining their business models to make these needed changes—this includes for example circular economy initiatives and B-Corps—but none of these changes are yet mainstream [1]. Recent studies have identified Sustainable Business Model (SBM) archetypes as a means to enable users to understand potential impacts of innovating in relation to different types of business models [2], and have also developed a new business model canvas that incorporates social and environmental layers, expanding on the original (economic) business model canvas [3]. However, none of these studies explain how companies that are in the process of transforming their traditional business models into more sustainable ones can use the current sustainability information they have, to advance to these innovative and more SBMs. Companies that are in this process usually have already identified the material issues for the stakeholders and for the company management. To help companies speed up their business model transformation, management needs a simple mechanism to link their company's materiality matrix (MM) to a SBM archetype in order to align and identify the innovation needed; and to facilitate their advance from a traditional business model canvas to a sustainable business model canvas. This will help transitioning companies make the necessary changes to their business models in a more practical and intuitive way.

In this context, and aligned with increasing attention and policy focus on global food security, the agri-food sector has a key role to play in sustainably producing and providing safe and affordable food for all. Within the agri-food sector, various scholars have logically highlighted the wine industry's inextricable connection with the core fundamentals of sustainability [4–6]. Furthermore, the industry is inherently linked to the terroir and other ecologically-related environmental aspects, which are directly associated with the product and its sustainability.

This study aims to understand the role the material issues identified in a company's materiality matrix (MM) play in identifying its SBM archetype as developed by Bocken et al. [2]; and the value creation process as proposed by the Triple Layered Business Model Canvas (TLBMC) developed by Joyce and Paquin [3].

Towards this end, a case study of an established major Chilean winery, Viña Concha y Toro S.A. (VCT) is undertaken to analyse and illustrate this holistic view of a company's value creation process. Currently, Chile ranks as the seventh largest wine producer, and is also the fourth largest exporter globally. Established in 1883, and spanning generations, the case company VCT is based in Chile, and is considered Latin America's largest winery. As one of Chile's oldest wineries, VCT is also one of the world's top ten wine exporters. The firm is an ideal candidate for research seeking case-specific, rich, and deep applied understanding into sustainability operationalisation; and the fact that the company is extensively internationalised, brings into play various cultural contexts including myriad stakeholder interaction and relational dependencies in its supply and value chains. Over the years, the company has instituted various sustainability initiatives, gradually evolving into a broader holistic commitment informing their strategy.

We undertake a detailed analysis of VCT's sustainability reports for the period 2017–2019. Furthermore, we also extensively interview the deputy sustainability manager of the company to garner deeper insights on the key sustainability aspects underpinning this study. We analyse the changes in VCT's material issues and any associated changes in their prioritisation, through the MM analyses between 2017 and 2019—to understand how these changes impact in the SBM archetype and value creation process.

Our study draws from stakeholder theory, given that this perspective associates value creation with and for stakeholders [7]. We contribute to SBM literature as we show how the MM, conceived by a multidimensional expression of materiality, relates to the SBM archetypes developed by Bocken et al. [2], and the Triple Layered Business Model Canvas (TLBMC) developed by Joyce and Paquin [3]. We also contribute to stakeholder theory in that we provide evidence that the value creation of a SBM is brought about by taking into account stakeholder demands. An effective SBM creates value for stakeholders aligned with their requirements, which in turn creates a strong link between the company and its stakeholders—probably stronger than in a common generic business model. Finally, this study offers the opportunity to understand how the wine industry is changing its sustainability material issue priorities. This also underlies the contribution of our study to stakeholder theory, showing the potential that stakeholders have—through the identification of material issues—to help a company transform its business model into a more sustainable one.

2. Literature Review

Though scholars may dispute extents of conceptual similarity or difference among fundamentals of Corporate Social Responsibility (CSR) and stakeholder theory [8]—mounting sensitivity to veritable environmental concerns and finite resources [9], combined with incessant visible cases of corporate malfeasance, and a post-financial crisis questioning of sustainability in employed capitalist ideals—has invariably seen key aspects of the two perspectives converge. Beyond public and societal demands, increasing policy and regulatory requirements justly see mounting pressure on businesses to bring stakeholder and sustainability responsibilities to the fore of their agendas. Further to maintaining awareness and striving for relational harmony among vested parties with at times inherently conflicting motivations, stakeholder theory "begins with the assumption that values are necessarily and explicitly a part of doing business" [10]. Managerial in its application, Freeman [11] consolidates positing that stakeholder theory essentially encapsulates two fundamental questions. The first, "What is the purpose of the firm?"—directs managers to understand and establish the shared sense of value they create, and what, through sustainable business enterprise draws its key stakeholders together. Moreover aligned with Porter's [12,13] strategic competitiveness-derived notion of shared value, "this propels the firm forward and allows it to generate outstanding performance, determined both in terms of its purpose and marketplace financial metrics" [10] (p. 364). The second question asks of management, "what responsibility do you owe to stakeholders?" This requires managers to articulate how they intend to conduct business specifically and operationally, and establish the kinds of relationships and rapport they need and want to create with their stakeholders to deliver on their purpose [10]. The latter in particular, inferring the need to visibly communicate, gauge and account for these relationships with stakeholders.

While keeping in mind the undisputed importance of shareholders and profits, critical in sustaining operations and growth—stakeholder theory underpins the need for managers to develop and nurture mutually sustaining relationships, inspire their stakeholders, and create communities where all parties contribute to deliver the value committed to by the firm [7,14,15]. Significantly, important profits become the result, rather than sole driver in the value creation process [8,16–18]. Given current realities, Freeman et al. [10] (p. 364) holistically observe that at the core lies the notion that "Economic value is created by people who voluntarily come together and cooperate to improve everyone's circumstance"—in itself an inferred requirement for sustainability. Stakeholder theory implies that stakeholders will support a company if they get value back in exchange [7,19], along the long-term cycle and mutual quest for sustainability.

2.1. Sustainable Business Models and Stakeholders

The origins of stakeholder theory predate the contemporary popular notion of business models in the literature. Initially a more nuanced view on capitalism, stakeholder theory emphasised the interconnectedness and relationships between a business, and essentially the entities it must, to greater or lesser extents, symbiotically interact with in order to sustain its operations and enterprise—namely, its customers, suppliers, employees, investors and communities among others—i.e., its stakeholders, as opposed to a quasi-sole focus on shareholder primacy. In 1984 Freeman [20] had consolidated various perspectives at the time, detailing his 'stakeholder theory of organisational management and business ethics'—at a time when business culture and common perception yet considered the notion of ethics and any corporate social responsibility beyond Milton Friedman's [21] paramount emphasis on shareholder's profits and adherence to laws and regulations—not comfortably reconcilable with the motives of business enterprise—put mildly. While Freeman [20] had coined the now classic definition of the concept: 'any group or individual who can affect or is affected by the achievement of the organisation's objective'—elusive universal consensus still sees scholars contest and debate defining 'stakeholders', such as 'who, and what counts' [22,23].

In this regard, today's heightened awareness, questions original notions of stakeholder theory—as societal, political and regulatory demands converge, perceiving broader holistic perspectives of stakeholders, and enterprises' obligations [22,24]—since businesses are now being actively seen as members of the societies within which they operate. This, essentially bolstered contemporary stakeholder perspectives, broadening the perceived remit of businesses' responsibilities, co-acknowledging and aligning both corporate and social intent [25]. Perhaps the recent declaration by Freeman [26], considered the father of stakeholder theory, effectively echoed what is now a generally acknowledged realisation across all stakeholder groups: that 'Managing Stakeholders' is the theme for the 21st century, and that the task of executives is to create as much value as possible for stakeholders without resorting to trade-offs. Great companies endure because they manage to get stakeholder interests aligned in the same direction. In such contexts, the capacity to 'endure' derives directly from sustainability.

Complementarily, business models are developed, configured and operationalised to create value [7,27–29]—with some form of value proposition at their core. While the conceptual notion of a business model is not new, 'business models' per se comparatively gained scholarly interest more recently [29]. Among various alternative models, Osterwalder and Pigneur's [30] business model canvas is considered among the most well known and extensively deployed by businesses—also equally acknowledged in academic circles. However, typical of non-sustainability-oriented business models, its emphasis is on unidirectional economic value—where the business creates value for the customer in exchange for financial economic value transacted for the business [7].

Drawing from the sustainability movement, and aligned with stakeholder theory, Stubbs and Cocklin [31] and Lozano [32] proposed that a SBM must consider all the stakeholders needs—incorporating social and environmental dimensions beyond economic imperatives. Stubbs and Cocklin [31] specifically underline the requirement for sustainable organisations to adopt a stakeholder rather than a shareholder view of the firm—highlighting that a company's longer term (and hence sustainable) success is inextricably related to the success of its stakeholders. In this line, these authors state that companies also need to treat nature as a stakeholder and promote environmental stewardship; from a holistic view. Accordingly, Upward and Jones [33] and Stubbs and Cocklin [31] probably make the first steps toward developing a SBM theoretical framework.

Concurrently, and drawing from an inherent need to account for value, the emergent and complementary 'triple bottom line' concept [34] inferred that the overall outcomes from a business model had to invariably also consider ecological and social, besides economic performance.

Guided by this, and building on the established and comprehensive yet easy to visualise and deploy Osterwalder and Pigneur [30] business model canvas, Joyce and Paquin [3] incorporated these two additional sustainability components and developed the TLBMC. Here, 'stakeholder' and 'lifecycle' perspectives respectively inform components on the two additional social and environmental canvas layers, expanding on the original economic business model canvas—seeking to account more fully for sustainability at strategic and operational levels. This represents an inside-out approach to analyse business model innovation. Given its foundation on the established and widely-used Osterwalder and Pigneur [30] canvas, coupled with its novel ease of use in visually enhancing analyses in conceptualising high-level sustainability-oriented innovation and operational strategies—we adopt this triple-layer framework for the purposes of our case investigation. As inferred earlier, sustainability scholars stress that the value of sustainability be necessarily shared among all actors, including "the natural environment and society as main actors" in order to be realised, extended and maintained over the longer-term [30,35]. In this regard, literature underlines that SBMs are necessarily holistic in their scope—where respective components are intimately interconnected through the stakeholders and explicable only by reference to the whole. Grigorescu et al. [36] point out that stakeholders are critically relevant in the SBM, playing an important role in consistently incorporating sustainability objectives into business models—concomitant longer-term collective outcomes cascading beyond the enterprise, to the national level.

In view of this, and the aforementioned core relevance of stakeholders in SBMs, the theoretical basis informing this research is stakeholder theory. On this foundation, Freudenreich et al. [7] propose a new approach, where business models create value, organise and facilitate the exchange of value creation with and for stakeholders, their proposed stakeholder value creation framework directly relates the organisation to its stakeholders (Figure 1), it shows this value creation in relation to the confluence from both the business model and stakeholder theory perspectives. Building on Freeman [11], these authors stress that the business model must consolidate value creation at the nexus of business model and stakeholder theory perspectives. They posit that in considering value creation, a SBM must answer: what and how sustainable value is created (deriving from traditional business model perspectives), whereas from the stakeholder theory perspective, the with and for whom that value is created, is addressed.

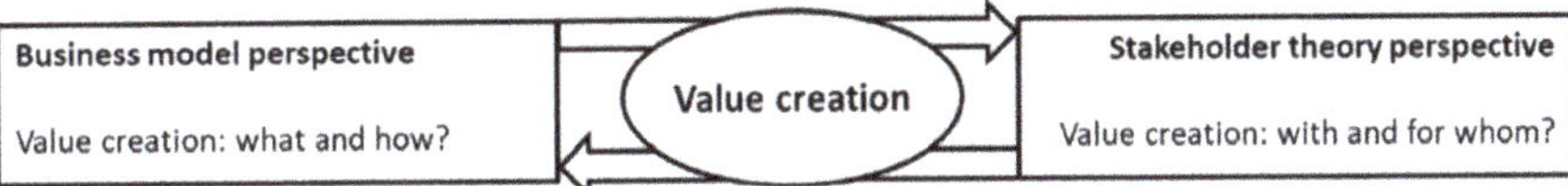

Figure 1. Business model and stakeholder theory perspectives on value creation.

In relation to the company, this stakeholder value creation framework establishes five stakeholder groups (societal, financial, employees, customers and business partners) with and through which value creation dimensions, activities and interactions take place [7]. These authors state that in this framework and within the joint creation processes, stakeholders are both (co)creators of value and receptors of the value created. Such mutually sustaining value generating dynamics across stakeholder categories have been empirically observed and also deemed beneficial and effective in agricultural contexts [9]. The business model archetypes identified by Bocken et al. [2] configures an outside-in approach to allow users to understand the potential impacts of innovating in relation to different types of business models. The archetypes are: maximise material and energy efficiency; create value from 'waste'; substitute with renewables and natural processes; deliver functionality rather than ownership; adopt a stewardship role; encourage sufficiency; repurpose the business for society/environment; and develop scale up solutions. The SBM archetypes describe groupings of mechanisms and solutions that may contribute to enhancing and building up the business model for sustainability; aiming at developing a common language useful in facilitating and enabling the development of sustainable business models in practice [2].

2.2. The Materiality Matrix

As acknowledged earlier, investing one's intent and attention on sustainability and talking about 'value' without the capacity to observe, gauge or assess any such initiative for the purposes of management and goal attainment is logically a moot point. This fundamental need (to gauge, measure and account for) becomes more critical (both internally and externally) when driven by mounting social (and therefore stakeholder) expectations and regulatory pressures transposed into evermore quantified obligations and compliance requirements. Business models at their core inherently infer and align with the need to analyse, assess and measure—given their strategic scope directly linked to prospective opportunities and performance. This need to measure and assess, had seen the development of Osterwalder and Pignur's [37] original business model canvas draw from Kaplan and Norton's [38,39] balanced scorecard. In itself a strategic management tool and framework, this balanced scorecard seeks to manage, assess and direct organisational performance on a broader set of factors deemed of strategic competitive importance, beyond the usual focus on financial metrics. Resulting from increasingly perceived needs to account for even broader stakeholder-related factors in a more focused manner, scholars further developed extended derivatives from Kaplan and Norton's original balanced scorecard. Significant interest in incorporating sustainability metrics for establishing and auditing business performance saw the development of the sustainability balanced scorecard [40]. See also, Hansen and Schaltegger [41]; Figge et al. [42]—which in turn, and to differing extents helped shape aspects of emerging SBMs. While the important capacity for management to account for, value and audit the linkage between strategic direction and goals set, and actual measurable progress or performance attained—is acknowledged as central to business models—it is however at times elusive or more challenging to quantify aspects of sustainability beyond generic inferences. This more so in the case of complex non-financial, qualitative measures. Addressing this particularly testing issue for SBMs, in their conceptual paper on assessing sustainability-oriented business models, Lüdeke-Freund et al. [40] (p. 169) highlight this SBM assessment gap and declare: "Whether and how 'sustainable business models' effectively support sustainable development is not just a matter of design, but also of the measurability and manageability of business model effects".

Originally derived from financial accounting and legal spheres [43], the concept of materiality highlights and discerns what is relevant and important. By extension and via application, materiality assessment is today also broadly adopted and directly linked to both CSR and sustainability performance—and, thus, invariably concerns stakeholders, given the usage of, and impact on resources and contingent effects on organisations' ecosystem realities. This linkage was arguably prompted by Starik in 1995 [30] who seriously asked: 'Should trees have managerial standing?'—and called out as a serious omission the non-recognition of nature as a stakeholder [44,45]. Addressing this persistent tendency for denying the environment stakeholder status, contemporary sustainability practice sought to transpose this into materiality assessment. For example, the international Association of Chartered Certified Accountants (ACCA), considered the global body for professional accountants, together with consulting and auditing firm KPMG and other associated environmental partners, staked their commitment in 'Identifying natural capital risk and materiality' [46]. Since its establishment in 2000, the Global Reporting Initiative (GRI) established an evolving portfolio of sustainability aspects that serve as an enterprise sustainability reporting guideline. Recognised and adopted worldwide across industries in the corporate world, the practicality and application of this assessment framework has also been acknowledged by scholars [40,47–50]. In its most recent iteration, the GRI reporting standards were explicit about the need to broaden consultation on aspects of stakeholder engagement. The GRI G4 guidelines further stress that reporting organisations should determine materiality and identify a process for accounting for such perspectives—including the interests of any stakeholders with whom the business may not be in constant or obvious dialogue. In this respect, it could be said that the materiality concept emerged as the most important element in the new GRI G4 guidelines on corporate sustainability reporting—especially, for instance, in the agricultural sector [43].

The GRI framework offers a sound guideline for sustainability reporting, and in the case of the G4 edition, specifically enhances this through the MM. This edition allows for a broad stakeholder-oriented approach in defining sustainability priorities, which, in our case, relate to the wine industry which forms part of the agri-food sector. In this regard, through the MM, the GRI's stakeholder approach is also useful for developing our research. The MM requires that the relevant sustainability aspects, from both the company's, and the stakeholder's perspectives, are juxtaposed—seeking to match and align both in the MM [40] (Figure 2).

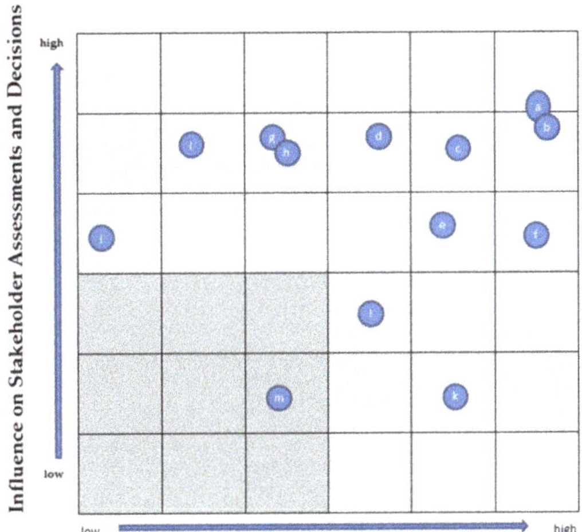

Figure 2. Illustration of a materiality matrix (MM).

Seeking more graduated assessment, each of a MM's material issues can also be identified with, and attributed a number to facilitate further evaluation—where higher values, indicate higher attributed priorities [51].

2.3. Material Issues in the Wine Industry

Ouvrard et al. [52] note that in the context of intense global competition, mounting societal expectations and market demands, wine producers and the broader industry ecosystem are generally very keen on environmentally friendly businesses; and sustainability and environmental issues tend to be reflected in their business models. Benson-Rea et al. [53] observe that in the New Zealand wine industry, multiple business models co-exist alongside each other. In wine production and distribution, topics related with environmental issues include land, water, energy, and chemical use, the generation and management of organic and inorganic waste streams, the production of greenhouse gas emissions, and the impact on ecosystems [54]. Olaru et al. [55] logically underline that sustainability of the wine industry involves environmental concerns in the grape production and processing systems.

In this industry the stakeholders' pressures drive sustainable practice [4,56]. With respect to stakeholder demands and requirements specifically associated with the agri-food sector, Dania et al. [57] establish stakeholders' sustainability requirements in agri-food supply chains across economic, environmental, and social dimensions (Figure 3).

2.4. Research Proposition

From the literature review in Section 2.1 we identified prior studies that have developed some solutions to help companies transition from a traditional business model to a sustainable business model. Specifically, Joyce and Paquin [3] developed the TLBMC to expand the original economic business model canvas allowing firms to account more fully for sustainability at strategic and operational levels showing the triple bottom line impact on stakeholders. Bocken et al. [2] identified business model archetypes that may contribute to building up the business model for sustainability and aim to develop a common language useful in accelerating the development of sustainable business models in practice. From the stakeholder theory perspective, stakeholders have a relevant role to play in the business model innovation process. Freudenreich et al. [7] propose a new approach, where business models create value, organise and facilitate the exchange of value creation with and for stakeholders—while Grigorescu et al. [36] point out that stakeholders are critically relevant in the SBM, playing an important role in consistently incorporating sustainability objectives into business models. However, none of these studies have discussed how the relevant sustainability aspects from both the company's and the stakeholder's perspectives represented in the MM (see Section 2.2), can be used by companies to apply the solutions developed in the abovementioned studies—to turn their current business model into a more sustainable one, and thus, to create sustainable value for their stakeholders and society as a whole. Based on this, we develop the following research proposition:

> Research proposition: The material issues identified in a company's materiality matrix (MM) are useful to align and determine the Sustainable Business Model archetype (SBM archetype) and the triple bottom line impact on stakeholders (TLBMC).

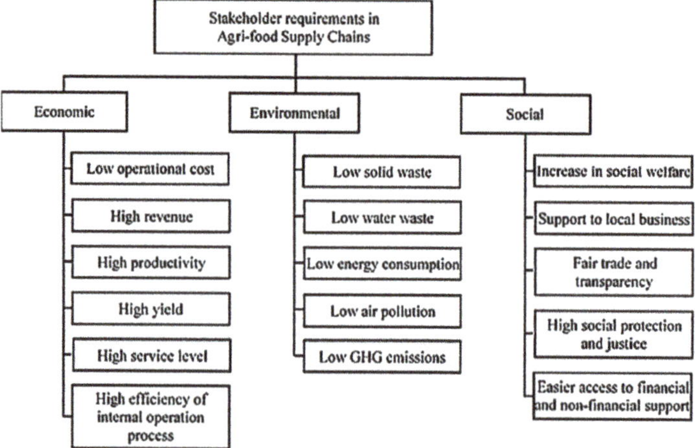

Figure 3. Stakeholder requirements in a sustainable agri-food supply chain.

Ensuing support for this proposition, should see companies able to utilise the MM they would have already prepared for sustainability reporting purposes, as a gateway to help transform their business model into a more sustainable one.

3. Materials and Methods

To understand the role that the MM plays in identifying the SBM archetype and value creation process in a SBM, we use an in-depth case study based on Viña Concha y Toro (VCT)—a well-established and internationalised wine grower and producer operating in a sector characterised by its inextricable link to elements fundamental to core aspects of sustainability. Such a case study approach allows rich contextually applied insights, and the analysis of empirical projects [58] that involve research and theory in the early or intermediate stages of development [59,60]. In this regard, such approaches have been effectively used in sustainability studies [61]. Case studies enable one to transform qualitative evidence into deductive research [60]. This methodology is used to gain an understanding of the processes and social interactions that develop in organizations in a specific historical context [62]. The objective of the case study was to extract information about: (1) VCT's sustainability approach; (2) VCT's materiality matrix; and (3) VCT's SBM elements. This is based on a thorough evaluation and content analysis of documentary evidence provided by VCT's extensive annual sustainability reports published on their website: Sustainability Report 2017 [63], Sustainability Report 2018 [64], and Sustainability Report 2019 [65].

Content analysis has been used to study a range of disclosure types in the accounting literature (e.g., [66,67]), and more specifically in this connection, one notes it is a common approach in CSR reporting [68].

The Sustainability Report 2017 [63] was assured by the external auditor provided by Deloitte, and was performed under the International Standard on Assurance Engagements (ISAE) 3000. ISAE 3000 is the assurance standard for non-financial information, and is issued by the International Federation of Accountants (IFAC). ISAE 3000 is usually applied for the audit of internal control, sustainability and compliance with laws and regulations. The sustainability reports for 2018 [64] and 2019 [65] were assured by the external auditor AENOR, who issues GRI standards certificates of compliance. The report review undertaken by the external auditor consisted in an enquiring process on different VCT units and management areas which had been involved in the development processes and drawing up of the report—as well as in the application of analytic procedures and checking tests. On the basis of procedures, the auditors state that nothing comes to their attention which causes them to conclude that the selected data for the sustainability reports has not been prepared in all material respects in accordance with the GRI reporting guidelines. Moreover, on the basis of validation from well-known and established international audit firms, we consider this information reliable for the purposes, scope and research objectives of our study.

The sustainability reports' content analysis is furthermore supported by our in-depth interviewing of VCT's deputy Sustainability Manager, providing further complementary qualitative and quantitative insights on detailed aspects beyond what was disclosed in the sustainability reports. The sustainability reports' content was thoroughly analysed, with a focus on the evolution of sustainability pillars and elements defined by the company between 2017 and 2019. In parallel, and aligned with our research objectives, analytic attention also converged on identifying the priority evolution of sustainability aspects, specifically the changes in material issues and the changes in their prioritisation, through the MM analysis. The first step in this study is to identify VCT's business model elements and identify the most relevant sustainability aspects through the period 2017–2019. The second step is to analyse the sustainability aspects prioritisation, in this sense the MM provides the sustainability priorities—matching the stakeholders' sustainability priorities with the company's sustainability priorities, in a matrix format [40]. The third step is to analyse possible matching between the material issues identified in the MM, and the established VCT sustainability pillars; in relation to ensuing elements consolidated in the TLBMC, and the SBM archetypes.

4. Results

4.1. VCT Sustainability Approach

"VCT's vision of sustainability is based on understanding that economic success goes hand in hand with caring for the environment, making rational use of natural resources, coupled with a commitment to people and the social sphere in which it operates. This virtuous circle is essential in the company's business model" [65] (p. 28). The definition of the objectives' content and strategic foci were based on their analysis and ensuing themes aligned with the winery's main stakeholders—identifying areas and issues requiring internal and/or external management to achieve strategic goals [65]. In 2018, VCT defined its 2022 corporate strategy, aiming at growth in business profitability and the creation of value based on the strategic pillars of excellence, sustainability and innovation; further including in 2019 the people pillar [65]. The components of VCT's strategic model incorporate the sustainability strategy into its core business: the production of high-quality wines. The sustainability strategy considers the product as the central element, and the strategic pillars emanate from and support this core element. In alignment, VCT's business model is articulated as follows: "The business model demands that the company participate actively in each of the stages of the value chain; vineyards, winemaking cellars, bottling plants and commercial offices, giving the company a vertical integration that assures the quality of each of their processes and of the final products" [64] (p. 18).

The winery defined its sustainability strategy around the following six strategic pillars, based on an analysis of the most relevant issues aligned with its key stakeholders. Each pillar's objectives contribute to fulfil VCT's vision:

1. Product: provide products of excellence that create the best experience for our customers.
2. Customers: create partnerships with our customers.
3. Supply Chain: be a partner for our suppliers.
4. People: have highly committed employees.
5. Society: create shared value for society.
6. Environment: be an example for the industry on environmental practices

To monitor the implementation of its sustainability strategy, VCT created a Sustainability Executive Committee involving leading executives that manage various pillars, the General Manager, and the Sustainable Development Area (led by VCT's deputy Sustainability Manager). In this way, sustainability became an essential element of the company, differentiating and positioning VCT as an exemplar for the industry in global markets. The company's Sustainability Strategy is aligned with the 2030 Agenda and the Sustainable Development Goals (SDGs) defined by the United Nations. While considering all the goals important and interconnected, the company nevertheless focuses efforts on those that are critical to its business and where they can have the greatest positive impact.

In 2012, VCT issued its first Sustainability Report prepared under GRI methodology. In this regard, in 2020, for the sixth consecutive year, VCT has been included in the Dow Jones Sustainability Index, an established international sustainability index assessing economic, social, and environmental aspects of a business, as well as corporate governance. Besides, the company has also been bestowed various awards associated with both sustainability as well as their wine brands.

The company defined eight categories of stakeholders, these categories were prioritised according to the stakeholder's degree of influence and interest in the organisation. The categories were classified as external and internal, according to the type of relationship they have with the company. VCT Internal stakeholders are the following: Employees; Shareholders; and Investors. VCT External stakeholders are the following: Suppliers; Communities; Society; Government and authorities; customers and the media. Moreover, VCT [65] (p. 10) declare: "The company seeks to encourage the engagement of all its stakeholders, with an emphasis on continuously promoting collaboration through various activities and communication channels where demands, opinions, concerns and suggestions can be expressed".

VCT's deputy Sustainability Manager states that the drivers that led the company in their sustainability initiative were: "1. The external driver: it came to the company around 2007 when the first formal requests for information regarding the company's sustainable management began, at that time, very influenced by the role that retail had taken. At that beginning, the responses that the company provided regarding the information requirements were rather informative and without compromising future performance regarding the different matters, given that the reported practices were only those that were implemented intuitively. The main concern of retail more than a decade ago was of an environmental nature, regarding the existence of analysis of impacts or minimal indicators. 2. The internal driver: when we realise that the company did not have a systematic management on the subject, the creation of a department in charge of proactively managing and promoting environmental and social issues within the company was formalised. In addition to formalisation in terms of functions, it is established that the Sustainable Development area operates in a transversal manner and acting as an internal facilitator. In addition, the management is formalised, through the generation of a Strategic Plan 2012–2015 in the first stage, which has renewed its continuity for the period 2015–2020".

4.2. VCT MM Evolution

The company, through surveys and interviews with employees, suppliers and other stakeholders, carried out [65] (p. 8): "a materiality analysis considering the results of the previous year and the structure of its Sustainability Strategy as a basis, in order to update it, adapting to the changes, trends and new challenges in the matter". The VCT MM represents the VCT prioritisation of material topics. Between 2017 and 2019, the material topics were analysed in order to represent its evolving prioritisation. Figure 4 shows that in 2019, the company focuses on four material themes which are distributed in three prioritised groups. These material themes are: water management; mitigation and adaptation to climate change; employees' well-being; and waste management and recycling.

VCT MM 2017 (3x3) 8 groups: 33 material themes	VCT MM 2018 (3x3) 6 groups: 40 material themes	VCT MM 2019 (4x4) 9 groups: 34 material themes
3.3.07 Water	3.3.02 Water Management	4.4 Water Management
3.3.06 Energy and Carbon Footprint	3.3.01 Energy	4.3.02 Mitigation and Adaptation to Climate Change
3.3.05 Waste	3.2.13 Emissions and Carbon Footprint	4.3.01 Employees Well-being
3.3.04 Health and security	3.2.12 Impact of Climate Change on the Business	3.4.01 Waste Management and Recycling
3.3.03 Remuneration	3.2.11 Research and Innovation	3.3.13 Energy (efficiency and use of renewable energy)
3.3.02 Welfare and Benefit	3.2.10 Working Conditions	3.3.12 Commitment to Sustainability
3.3.01 Community management	3.2.09 Communication of Corporate and Sustainability Strategy	3.3.11 Biodiversity and Soil Care
3.2.08 Impacts of Climate Change on the business	3.2.08 Waste Management and Recycling	3.3.10 Human Rights
3.2.07 Biodiversity	3.2.07 Transparency and information to customers	3.3.09 Human Capital Development
3.2.06 Training and Knowledge Center	3.2.06 Ethics and Anticorruption	3.3.08 Diversity and Equal Opportunities
3.2.05 Supplier Development	3.2.05 Health and Safety	3.3.07 Legal Compliance
3.2.04 Innovation	3.2.04 Human Rights	3.3.06 Responsible Sourcing
3.2.03 Quality management	3.2.03 Product Innovation	3.3.05 Ethics and Anti-corruption
3.2.02 Products with positive impacts	3.2.02 Regulatory Compliance	3.3.04 Innovation, Research and New Technologies
3.2.01 Ethics and anti-corruption	3.2.01 Promoting Sustainability in the Supply Chain	3.3.03 Support for Local Development
3.1.01 Social impact of products	3.1.06 Product Quality and Safety	3.3.02 Promotion of Sustainability in the Supply Chain
2.3.01 Management and Evaluation of suppliers	3.1.05 Corporate Culture	3.3.01 Health and Safety
2.2.10 Pesticides and Fertilizers	3.1.04 Remuneration, Welfare and Benefits	3.2.06 Supply Management (grapes and materials)
2.2.09 Working conditions	3.1.03 Engagement and Working Environment	3.2.05 Product Quality and Safety Managemen
2.2.08 Career development	3.1.02 Initiatives that promote the protection of the environment with communities and employees	3.2.04 Internal Culture
2.2.07 Relations and work climate	3.1.01 Carrer Development	3.2.03 Risk Management
2.2.06 Diversity and equal opportunities	2.2.09 Suppliers Management and Evaluation	3.2.02 Corporate Strategy
2.2.05 Materials	2.2.08 Contribution to SDG's	3.2.01 Excellence in Operations
2.2.04 Promotion of sustainability in the supply chain	2.2.07 Community Management	2.3.03 Innovation and New Products
2.2.03 Profitability and economic value	2.2.06 Diversity and Equal Opportunities	2.3.02 Customer Satisfaction
2.2.02 Human Rights	2.2.05 Quality Management	2.3.01 Marketing and Responsible Drinking
2.2.01 Responsible Marketing	2.2.04 Relations and Customers Satisfaction	2.2.05 Profitability and Economic Indicators + Economic Performance
2.1.04 Supply and logistics management	2.2.03 Responsible Consumption	2.2.04 Awards, Recognitions and Strategic Alliances
2.1.03 Customer Satisfaction	2.2.02 Responsible Marketing	2.2.03 Information Security
2.1.02 Transparency and information to customer	2.2.01 Contribution to the Wine Industry	2.2.02 Cerfications
2.1.01 Fulfillment	2.1.09 Vineyards Management	2.2.01 Transparency and Customer Information
1.2.01 Vineyard management	2.1.08 Profitability and Economic Indicators	2.1.02 Origins and Portfolio (Family of Wineries)
1.1.01 Dissemination and Promotion of sustainability	2.1.07 Certifications	2.1.01 Efficiency in Distribution
	2.1.06 Training and Knowledge Center	1.2.01 Social Initiatives and Volunteering
	2.1.05 Biodiversity	
	2.1.04 Information Security	
	2.1.03 Brands Management	
	2.1.02 Social Initiatives and Volunteering	
	2.1.01 Materials and Supplies	
	1.1.01 Awards and Recognitions	

Figure 4. Viña Concha y Toro S.A. (VCT) 2017, 2018, and 2019 MM prioritisation of material themes.

The materiality process carried out by the company each year considers the results of the previous year, the structure of the sustainability strategy as a basis, and the necessary updates to adapt to changes, trends and new challenges. The VCT prioritisation process includes surveys and interviews with stakeholders, a review of the industry's sustainability context, and the gathering of internal information. Figure 4 represents VCT's materiality matrices for the years 2017, 2018, and 2019. Figure 4 was consolidated in line with the prioritisation of the material topics indicated by VCT each year in its materiality matrices published in its 2017, 2018, and 2019 sustainability reports. We assigned a number to each material topic, the higher the value, the higher the priority [51]. The numbers were assigned with the following logic: the first number indicates the priority level for the company, the second the priority level for the stakeholders and the third the priority of the item in the corresponding quadrant.

The company defined thirty-three material topics distributed in eight groups in 2017; forty material topics distributed in six groups in 2018; and thirty-four material topics distributed in nine groups in 2019 (Figure 4). Although the number of material topics and groups during 2019 is very similar to 2017, the difference lies in the focus of the topics distributed in their top three priority groups. In order to carry out this targeting, the company expanded its MM from 3×3 to a 4×4 matrix in 2019.

When comparing the MM between 2017 and 2019, the trend of the materiality process was to focus on the most relevant material issues, observing, during 2019, in the first three priority groups, the four material topics most relevant to the company (Table 1).

Table 1. VCT Number of material themes between 2017 and 2019, in the first three groups.

Material Themes in MM	2017	2018	2019
First prioritisation	6	2	1
Second prioritisation	8	13	2
Third prioritisation	1	6	1

During 2019, only four material topics (water management, mitigation and adaptation to climate change, employee wellbeing, and waste management and recycling) were concentrated in the three highest priority groups. On the other hand, in 2017, there were sixteen material topics in the top three priority groups, and in 2018, there were twenty-one. Figure 5 shows that water management represents VCT's number one priority. The company's highest priority issue on sustainability is water management (Figure 5). The table also shows that the highest priority area contained seven material topics in 2017, two topics in 2018 and only one in 2019, showing a focused strategy on water management. The changes made by the company between 2017 and 2019, in terms of prioritising of material issues, generated a targeting of VCT priorities.

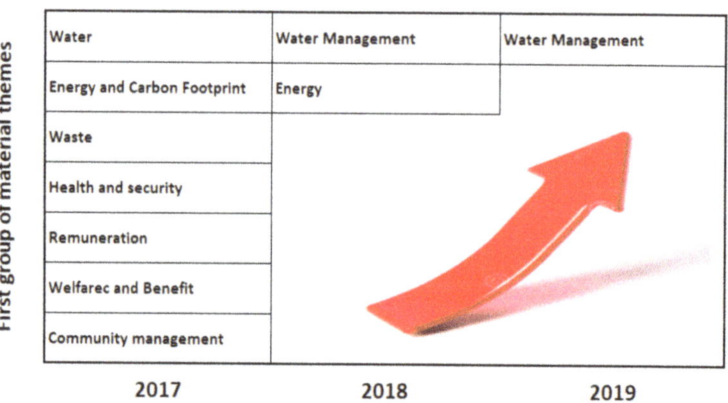

Figure 5. VCT material themes, prioritisation of first group tendency.

4.3. VCT SBM Elements

We defined the elements of the company's sustainable business model as those elements that are measured by the company in the main elements reported by VCT in each pillar of its sustainability strategy. The main elements reported by the company in each pillar in its annual sustainability reports were considered by this study as elements of the VCT SBM. In Figure 6, we show the VCT SBM elements between 2017 and 2019.

	2017 Pillars & Elements	2018 Pillars & Elements	2019 Pillar & Elements
	Environment Pillar 2017	**Enviromental Pillar 2018**	**Enviromental Pillar 2019**
Elements	Water	Water	Water
	Energy	Energy	Energy
	Biodiversity	Biodiversity	Biodiversity
	Waste	Waste	Circular Economy
	Climate change	Climate change	Climate change
	Supply Chain Pillar 2017	**Supply Chain Pillar 2018**	**Supply Chain Pillar 2019**
Elements	Responsible Supply Chain	Responsible Supply Chain	Responsible sourcing
	Carbon Footprint	Sustainability Index	Sustainability index
	Packaging	Sustainable packaging	Sustainable packaging
		Packaging carbon footprint	Packaging carbon footprint
	Product Pillar 2017	**Product Pillar 2018**	**Product Pillar 2019**
Elements	Innovation	Innovation	Innovation
	Quality	Quality	Quality
	Sustainable atributes	Sustainable attributes	Sustainable attributes
	Responsible drinking	Responsible drinking	Responsible drinking
	Customers Pillar 2017	**Customers Pillar 2018**	**Customers Pillar 2019**
Elements	Efficiency in logistics costs	Efficiency in logistics costs	Efficiency in logistics costs
	Efficiency in CO2 emissions	Efficiency in CO2 emissions	Efficiency in CO2 emissions
	Integral customers	Integral customers	Integral customers
	People Pillar 2017	**People Pillar 2018**	**People Pillar 2019**
Elements	Career development	Career development	Career development
	Engagement	Engagement	Engagement
	Knowledge center	Knowledge center	Training
	Ethical management	Ethical management	Ethical management
	Society Pillar 2017	**Society Pillar 2018**	**Society Pillar 2019**
Elements	Productive alliances	Productive alliances	Productive alliances
	Extension for grape growers	Extension for growers	Extension for growers
	Communities	Communities	Communities
	Education	Education	Education
		Entrepreneurship	Entrepreneurship

Figure 6. The VCT Sustainable Business Model (SBM) elements for the period 2017–2019.

The evolution of the elements of each sustainability pillar of the company between 2017 and 2019 are analysed and outlined below:

- In the Environment Pillar the evolution of its elements is related to the incorporation of the element of circular economy. The "waste" element was replaced by the "circular economy" element. The Environment Pillar elements that remain in 2019 are: water, energy, biodiversity and climate change.
- In the Supply Chain Pillar, the evolution of its elements is related to: first, the replacement of the elements "Carbon Footprint" by "Sustainability Index" and "Packaging" by "Sustainable packaging", both changes were made by the company in 2018. Second, the incorporation of a new element called "Packaging carbon footprint", a change made in 2018 and third, the Responsible Supply Chain element was replaced by the Responsible Sourcing element.
- In the Product Pillar, there are no changes related to its elements during the period, the elements are the following: innovation, quality, sustainable attributes, and responsible drinking.

- In the Customers Pillar. there are no changes related to its elements during the period. In the 2017–2019 period, the elements are the following: efficiency in logistics costs, efficiency in CO_2 emissions and integral customers.
- In the People Pillar, the evolution of its elements is related to the replacement of the element "Knowledge Center" by "Training" in 2019. The elements of the People Pillar that remain unchanged in the period 2017–2019 are the following: career development, engagement and ethical management.
- In the Society Pillar, the evolution of its elements is related to the incorporation of a new element, called "Entrepreneurship". The elements of the Society Pillar that remain unchanged in the period 2017-2019 are the following: productive alliances, extension for grape growers, communities and education (training).

4.4. Developing the Environmental and Social Canvas Layers for VCT

Using the elements of the VCT SBM 2019 described above, we developed the 'environmental life cycle' and the 'social stakeholder' layers of the TLBMC, as can be seen in Figures 7 and 8. The criteria and method used was to relate the elements of the VCT SBM with the framework in the form of the TLBMC developed by Joyce and Paquin [3]. We developed only the environmental and social layers of the canvas, as the elements of the VCT SBM are concentrated on these two topics.

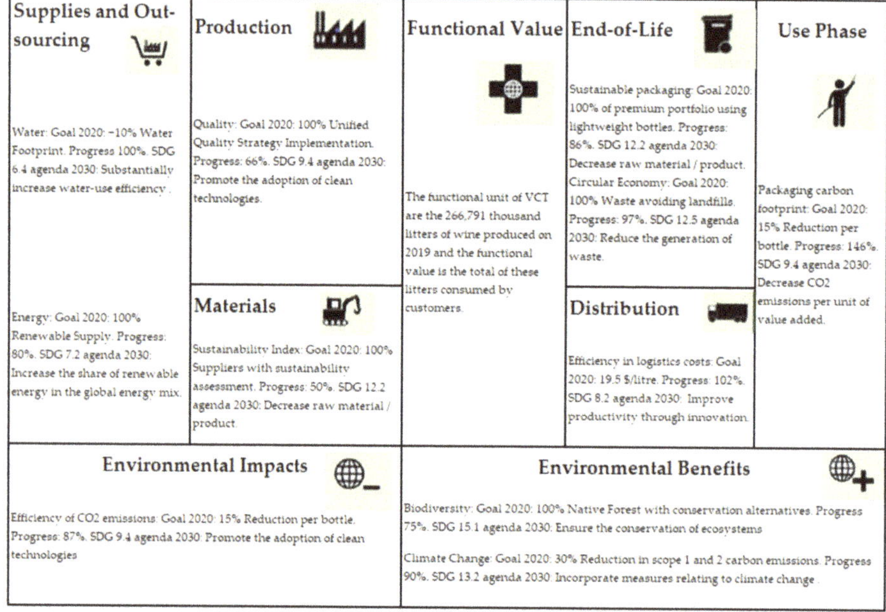

Figure 7. The environmental life cycle layer of the Triple Layered Business Model Canvas (TLBMC), 2019 VCT.

The TLBMC built for this case study, specifically the 'environmental life cycle' layer and the 'social stakeholder' layer of the TLBMC, allows us: first, to identify and establish a comprehensive vision of the elements of the company's sustainable business model; second, to specify the actions carried out by the company in terms of social and environmental sustainability; third, to have a holistic vision of the company's SBM showing the different types of value creation, in terms of both social and environmental sustainability; and fourth, to enable the integration of the different types of value creation, in terms of both social and environmental sustainability. Seeing how the overlaid SBM elements from the different strategic sustainability pillars defined by VCT match the different components of the SBM canvas' environmental and social layers, provided for useful and interesting insights directly aligned with our research objectives.

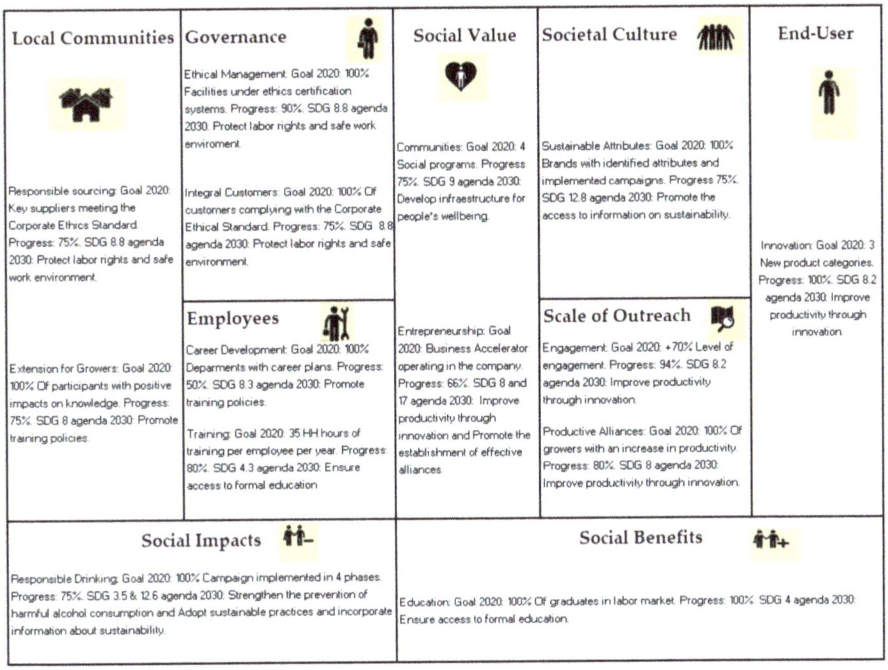

Figure 8. The social stakeholder layer of the TLBMC, 2019 VCT.

4.5. *Aligning the MM with the TLBMC and the SBM Archetypes*

To test our research proposition, we relate the elements of the 2019 VCT SBM (that were previously matched with, and consolidated in the TLBMC) with the material issues defined in VCT's 2019 MM. We do this in order to visualise how the company responds and adjusts its business model in relation to the sustainability issues included in the MM developed—consolidating and considering the importance of the material issues defined by both the stakeholders and VCT (Figure 9).

Figure 9 shows that all elements of the VCT TLBMC respond to the material issues identified in the company's MM. In some cases, each element of the SBM responds to more than one of the material issues identified. In the 'environmental life cycle' and 'social stakeholder' layers of the triple layered business model canvas, it is possible to observe the case organisation's progress towards the achievement of the goals related to each element of its SBM (Figures 7 and 8).

The goals linked to the four more relevant material issues and its level of achievement are the following:

- Water management: 10% reduction of water footprint (100%);
- Mitigation and adaptation to climate change: 30% reduction in scope 1 and 2 (90%);
- Employee wellbeing: 100% departments with career plans (50%);
- Waste management and recycling: 100% waste avoiding landfills (97%).

Finally, we link each of the material issues included in VCT's MM with the SBM archetypes (Figure 9). Out of the eight SBM archetypes, all the material issues included in the MM are related to four of them:

- Technological—maximise material and energy efficiency;
- Technological—substitute with renewables and natural processes;

- Social—adopt a stewardship role;
- Technological—create value from waste.

SBM Archetype (Outside-in approach)	2019 VCT MM material themes (relevance for VCT & Stakeholders)	2019 VCT MM material theme priority group	2018 VCT MM material theme priority group	2017 VCT MM material theme priority group	VCT Strategic Pillar	2019 VCT SBM Elements	SBM Canvas Layer (Inside-out approach)
Technological - Maximise material and energy efficiency	4.4 Water Management	1	1	1	Environmental	Water	Environmental - Supplies and outsourcing
Technological - Substitute with renewables and natural processes	4.3.02 Mitigation and Adaptation to Climate Change	2	2	2	Environmental	Climate Change	Environmental - Environmental benefits
Social - Adopt a stewarship role	4.3.01 Employees well-beig & 3.3.01 Health and Safety	2	2	1	People	Career Development	Social - Employees
Technological - Create value from waste	3.4.01 Waste Management and Recycling	3	2	1	Environmental	Circular Economy	Environmental - End of life
Technological - Substitute with renewables and natural processes	3.3.13 Energy (efficiency and use of renewable energy)	4	2	1	Environmental	Energy	Environmental - Supplies and outsourcing
Technological - Substitute with renewables and natural processes	3.3.12 Commitment to Sustainability	4	2	8	Supply Chain	Sustainability index	Economic - Value proposition
Social - Adopt a stewarship role	3.3.11 Biodiversity and Soil Care	4	5	2	Environmental	Biodiversity	Environmental - Environmental benefits
Social - Adopt a stewarship role	3.3.10 Human Rights & 3.2.03 Risk Management & 3.3.07 Legal Compliance & 3.3.05 Ethics and Anti-corruption & 2.2.03 Information Security	4	2	2	People	Ethical Management	Social - Governance
Social - Adopt a stewarship role	3.3.09 Human Capital Development	4	3	5	People	Career Development	Social - Employees
Social - Adopt a stewarship role	3.3.08 Diversity and Equal Opportunities	4	4	5	Society	Education	Social - Social benefits
Social - Adopt a stewarship role	3.3.06 Responsible Sourcing	4	5	5	Supply Chain	Responsible sourcing	Economic - Resources
Technological - Maximise material and energy efficiency	3.3.04 Innovation, Research and New Technologies & 2.3.03 Innovation and New Products	4	2	2	Product	Innovation	Economic - Activities
Social - Adopt a stewarship role	3.3.03 Support for Local Development	4	5	-	Society	Entrepreneurshi	Social - Social value
Social - Adopt a stewarship role	3.3.02 Promotion of Sustainability in the Supply Chain	4	2	-	Supply Chain	Packaging carbon footprint	Economic - Partners
Social - Adopt a stewarship role	3.2.06 Supply Management (grapes and materials)	5	4	6	Society	Extension for Growers	Social - Local communities
Social - Adopt a stewarship role	3.2.05 Product Quality and Safety Management & 3.2.02 Corporate Strategy	5	3	-	Product	Quality	Economic - Value proposition
Social - Adopt a stewarship role	3.2.04 Internal Culture	5	4	-	People	Training	Social - Employees
Social - Adopt a stewarship role	3.2.01 Excellence in Operations	5	4	-	Customers	Efficiency of CO2 Emissions	Economic - Activities
Social - Adopt a stewarship role	2.3.02 Customer Satisfaction	6	4	6	Supply Chain	Sustainable packaging	Economic - Customer relationship
Social - Adopt a stewarship role	2.3.01 Marketing and Responsible Drinking	6	4	5	Product	Responsible Drinking	Social - Social impacts
Social - Adopt a stewarship role	2.2.05 Protability and Economic Indicators Economic Performance	7	5	5	Product	Sustainable attributes	Economic - Revenues & Costs
Social - Adopt a stewarship role	2.2.04 Awards, Recognitions and Strategic Alliances & 2.1.02 Origins and Portfolio (Family of Wineries)	7	6	-	Society	Productive Alliances	Social - Scale of outreach
Technological - Substitute with renewables and natural processes	2.2.02 Certifications	7	5		Product	Sustainable Attributes	Economic - Resources
Social - Adopt a stewarship role	2.2.01 Transparency and Customer Information	7	4	6	Customers	Integral Customers	Economic - Customer relationship
Technological - Maximise material and energy efficiency	2.1.1 Efficiency in Distribution	8	4	-	Customers	Efficiency in logistics costs	Economic - Channels
Social - Adopt a stewarship role	1.2.01 Social Initiatives and Volunteering	9	5	-	Society	Communities	Social - Social value

Figure 9. The 2019 VCT MM linked to SBM canvas elements and SBM archetypes.

Interestingly, the four most relevant material issues are linked to these four SBM archetypes.

As shown in Figure 9, the MM is useful to, both, understand how the company creates value for stakeholders as shown in the TLBMC; as well as, to identify the SBM archetype that may contribute to building up the business model for sustainability. These ensuing results support our research proposition.

These empirical findings were consolidated and supplemented by complementary insights garnered by means of interviews with VCT's deputy Sustainability Manager, seeking support for the aligned SBM archetypes we identify for VCT. After understanding the different SBM archetypes, the deputy Sustainability Manager stated that there is no single SBM archetype that identifies VCT, but there are four that best encompass VCT's strategy, and also provided us with the rationale for that statement (using the methodology developed by Bocken et al. [2] we provided). The four SBM archetypes the manager identified for VCT, and the rationale provided for each respective choice, is shown below:

1. SBM archetype 1: maximise material and energy efficiency:
   a. Value proposition: packaging reduction; introduction of a light bottle with 13% less weight and therefore a reduction in the generation of waste and reduction of emissions in transportation and processing.
   b. Value creation and delivery: VCT worked with Cristalerías de Chile for the development of the new light bottle, the eco-glass format, which became a standard for the Chilean wine industry.
   c. Value capture: the light bottle provided for cost savings related to the main input of VCT's operations.
2. SBM archetype 2: create value from 'waste':
   a. Value proposition: currently 98% of waste is recycled, reused or recovered. Organic waste is used to generate compost that is applied again to the earth due to its high organic content, which helps to increase the health and productivity of the soils. VCT is moving towards 100% of waste destined for recycling, reuse or recovery.
   b. Value creation and delivery: VCT has different alliances for each type of waste to be recovered.
   c. Value capture: through its circular economy initiatives VCT generates savings for transport and disposal of waste, and the sale of waste. For waste that is generated on a smaller scale, alternatives for use are sought.
3. SBM archetype 3: substitute with renewable and natural processes:
   a. Value proposition: VCT has Initiatives to incorporate renewable energy. The company is moving towards a 100% renewable energy supply in all its facilities.
   b. Value creation and delivery: in order to communicate this attribute to its consumers, VCT generated a joint project with CRS (Centre for Resource Solutions) to bring to Chile the Green-e renewable energy certification standard, which enables the use of a seal on the product to promote and communicate recognition the said attribute.
   c. Value capture: the use of renewable energies has meant lower energy costs and a reduced carbon footprint. Through product labelling, VCT communicates this directly to consumers in the most receptive markets—emphasising the sustainable attributes of its products.
4. SBM archetype 4: adopt a stewardship role:
   a. Value proposition: application of ethical standards in the supply chain, through VCT's established responsible sourcing program.
   b. Value creation and delivery: through VCT certification of the Sustainability Code of Wines of Chile (Vinos de Chile), environmental and social aspects are worked upon through collaboration with grape suppliers, focusing on agricultural practices.
   c. Value capture: through supply chain programs, VCT has achieved and enjoys suppliers' loyalty. There are different types of programs depending on the provider segment. Supply chain programs are in place to enhance and advance suppliers' quality, productivity, and sustainability. This generates, promotes and fosters suppliers that operate in a coordinated manner with the organisation, improving sustainability and response rates.

It is interesting to note that the self-perception of VCT regarding the four SBM archetypes coincides with the same four SBM archetypes that—based on our analysis—we linked with the material issues included in the MM. This confirms that the linkage we established is aligned and appropriate, confirming the approach we propose and the usefulness of the MM in this endeavour.

From joint analysis we observed the following: first, the VCT SBM answers the "what and how value is created" questions from the sustainable business model theory perspective in the company, in an inside-out approach (Figures 7 and 8). Second, the VCT SBM also answers the "with and for whom the value is created" questions from the stakeholder theory perspective (Figure 9—because each of the SBM elements in the canvas are related to the material issues included in the MM). Third, water management represents the number one sustainability priority of the company (Figure 5), while the top four priorities are complemented in 2019 with two more environmental issues (mitigation and adaptation to climate change, and waste management); and a social issue (employee wellbeing)—as we indicated in Figure 4. Fourth, the material issues included in the MM are all linked to a different SBM archetype, and these archetypes are also linked to environmental and social topics, from an outside-in approach (Figure 9). Finally, the stakeholders' requirements in a sustainable agri-food supply chain as stated by Dania et al. [57] are related to environmental and social topics which are met by VCT, with the exception of the issue related to "easier access to financial and non-financial support", which is not explicitly specified in the company's sustainability reports.

5. Discussion and Conclusions

Previous studies have developed methodologies and solutions to help companies transform their business model into a more sustainable one, as is the case of the SBM archetypes [2] and the TLBMC [3]; but none of these studies explored how the materiality matrix could be a tool to help companies advance towards a sustainable business model. Our study is motivated by increasing expectations and the urgent need for companies to transform their business by introducing innovation to their business models in order to conduct business in a more sustainable way—seeking maximisation not only of financial and economic value, but also social and environmental value. Drawing from stakeholder theory, this study's research objective is to understand the role that the material issues identified in a company's MM play in identifying its SBM archetype and its value creation process as proposed by the TLBMC.

To test and establish support for this study's research proposition, we use an in-depth case study focused on Viña Concha y Toro, a world leading winery based in Chile. VCT conduct business in the agri-food industry, a sector of interest as these companies need to play a significant role in the 2030 Agenda since they are directly linked to SDG 12—'Responsible consumption and production', and SDG 2—'Zero hunger'.

This paper provides an approach, through the use of the MM, for linking the theoretical concept of the SBM archetype that aligns and refers to business model innovation, to the SBM elements represented in a TLBMC. Based on our results, we conclude that the MM has the potential to help companies identify the SBM archetype relevant to transform their traditional business model into a more holistically sustainable business model; and to also better enable an understanding of the dynamics that create triple bottom line impact on their stakeholders. This study proposes a tool that companies can use to transform their business model, and to advance toward more comprehensive strategic and operational sustainability using information from the MM they typically construct during their sustainability report preparation process.

We contribute to the SBM literature as we show how the MM, conceived by a multidimensional expression of materiality, relates to the SBM archetypes developed by Bocken et al. [2], and the TLBMC developed by Joyce and Paquin [3]. This also underlies the contribution of our study to stakeholder theory, showing the potential that stakeholders have—through the identification of material issues—to transform the firm's business model into one of sustainability. This study is novel in linking these three concepts to propose a useful tool for companies to advance in their sustainability journey. A strength of the methodology employed in this study is that based on content analysis, in that we gather the audited and documented information required to develop the basis of this research, and we in turn additionally validate and confirm our observations with complementary rich insights from in-depth interviewing with the senior sustainability management of Viña Concha y Toro, a recognised world leader in sustainability, ranked in the 2020 Dow Jones Sustainability Index.

Firms can use the approach we propose with the MM to understand how their stakeholders' view of material issues in the environmental, social and economic/governance perspectives influence both their value creation process, and the triple bottom line impact on stakeholders through shaping and informing their SBM elements. This will help companies to incorporate in their current business models the sustainability issues that progressively matter most to their stakeholders over time—and hence, enhance their clarity of vision and alignment in turning their business models into more comprehensively sustainable ones. This is an internal transformation of the current business model elements produced by the stakeholders' influence. Additionally, we propose to use the MM to identify the more suitable SBM archetypes, or a combination of SBM archetypes, that will allow the company to explore the potential impacts of innovating towards different types of business models. Hence, we identify the MM as the gateway for companies to innovate and develop a business model that allows them to deliver sustainable value to their stakeholders.

Additionally, and in consolidation, the purpose of understanding the role that the MM plays in shaping the SBM elements of the company and SBM archetypes, is also particularly relevant in current times, where new players expect to enter into the sustainability standards issuers' arena. This, more specifically given the International Financial Reporting Standards (IFRS) foundation's proposal seeking to impose a (simplified) single view of materiality more closely linked to the financial materiality view—in contrast to the multidimensional definition of materiality proposed by the GRI framework. In this regard, this study furthermore shows the enhanced relevance that the multidimensional definition of materiality represented in a MM has in: shaping the dynamics of a SBM from an outside-in approach; establishing association with the SBM archetypes developed by Bocken et al. [2]; and from an inside-out approach, articulating the value created, as in the TLBMC developed by Joyce and Paquin [3]. The simplified view of materiality, focused on the enterprise value creation process to shareholders, is a step back in the study of SBM from the holistic approach provided by the multidimensional view of materiality, which focuses on the organization's significant impact on the triple bottom line to a wider range of stakeholders [69].

That said, there are potential limitations to this proposed use of the MM. Firstly, the linkage of the material themes to the SBM elements in the canvas, and to the SBM archetypes is reflective, based on the current business model canvas and SBM archetypes. This analysis should be revisited periodically to identify new synergies. Secondly, the industry sector should be considered when using the MM for this purpose, as the SBM archetypes are related to different groups of business model innovations (technological, social and organisational), and each of them may be more suited to specific industries. Our research is based on a case study in the winery industry, other industries could pose different complexities.

Our study is timely as the issuers of sustainability standards and metrics are entering into a process of mergers, and in this regard new players will as expected emerge. The concept of materiality is one of the most relevant to be considered by companies when issuing a sustainability report, and we show that its multidimensional definition perspective should prevail due to its comprehensive nature and potential to promote business model innovation. Our findings and results create a straightforward methodology for companies to use in order to incorporate innovation and transform their business models towards sustainability. It helps sustainability standards issuers and financial reporting standards issuers to understand the link between materiality and the value creation process; as well as the triple bottom line impact on strategic operations through the dynamics in the SBM.

Our study provides a tool that the leading firm of a value chain can use to coordinate and require other members of the value chain to apply, with the purpose of identifying collaboration opportunities to align and comprehensively strengthen the sustainability of the value chain, by finding and establishing mutually reinforcing complementarities. This study also opens avenues for future research as we need to better understand how the multidimensional concept of materiality impacts the quantified figures of financial statements as a result of the dynamics in the company's value creation process—and more specifically, how these dynamics generate triple bottom line impact on stakeholders.

Author Contributions: Conceptualization: V.V.G.-W., N.G.; N.P.M.; S.L.G.-W. Methodology: V.V.G.-W.; N.G.; S.L.G.-W. Investigation: V.V.G.-W., N.G.; N.P.M.; S.L.G.-W. Writing—Original draft preparation: V.V.G.-W., N.G.; N.P.M.; S.L.G.-W. Writing—review and editing: N.P.M. Project administration: V.V.G.-W. All authors have read and agreed to the published version of the manuscript.

Funding: Funded by the **Universidad de La Frontera**, Project DI20-0087.

Conflicts of Interest: The authors declare no conflict of interest.

References

1. World Resources Institute. The Elephant in the Boardroom: Why Unchecked Consumption is not an Option in Tomorrow's Markets. Available online: https://www.wri.org/publication/elephant-in-the-boardroom (accessed on 8 January 2021).
2. Bocken, N.M.; Short, S.W.; Rana, P.; Evans, S. A literature and practice review to develop sustainable business model archetypes. *J. Clean. Prod.* **2014**, *65*, 42–56. [CrossRef]
3. Joyce, A.; Paquin, R.L. The triple layered business model canvas: A tool to design more sustainable business models. *J. Clean. Prod.* **2016**, *135*, 1474–1486. [CrossRef]
4. Gilinsky, A.; Newton, S.K.; Vega, R.F. Sustainability in the global wine industry: Concepts and cases. *Agric. Agric. Sci. Procedia* **2016**, *8*, 37–49. [CrossRef]
5. Maicas, S.; Mateo, J.J. Sustainability of wine production. *Sustainability* **2020**, *12*, 559. [CrossRef]
6. Santini, C.; Cavicchi, A.; Casini, L. Sustainability in the wine industry: Key questions and research trends a. *Agric. Food Econ.* **2013**, *1*, 9–22. [CrossRef]
7. Freudenreich, B.; Lüdeke-Freund, F.; Schaltegger, S. A Stakeholder's theory Perspective on Business Models: Value Creation for Sustainability. *J. Bus. Ethics* **2020**, *166*, 3–18. [CrossRef]
8. Brown, J.A.; Forster, W.R. CSR and Stakeholder theory: A Tale of Adam Smith. *J. Bus. Ethics* **2013**, *112*, 301–312. [CrossRef]
9. Savarese, M.; Chamberlain, K.; Graffigna, G. Co-Creating Value in Sustainable and Alternative Food Networks: The Case of Community Supported Agriculture in New Zealand. *Sustainability* **2020**, *12*, 1252. [CrossRef]
10. Freeman, R.E.; Wicks, A.C.; Parmar, B. Stakeholder theory and "the corporate objective revisited". *Organ. Sci.* **2004**, *15*, 364–369. [CrossRef]
11. Freeman, R.E. The politics of stakeholder theory: Some future directions. *Bus. Ethics Q.* **1994**, *4*, 409–421. [CrossRef]
12. Porter, M.E.; Kramer, M.R. The link between competitive advantage and corporate social responsibility. *Harv. Bus. Rev.* **2006**, *84*, 78–92. [PubMed]
13. Porter, M.E.; Kramer, M.R. Creating Shared Value. *Harv. Bus. Rev.* **2011**, *89*, 62–77.
14. O'Brien, I.M.; Jarvis, W.; Soutar, G.; Ouschan, R. Co-creating a CSR Strategy with Customers to Deliver Greater Value. In *Disciplining the Undisciplined? CSR, Sustainability, Ethics & Governance*, 1st ed.; Brueckner, M., Spencer, R., Paull, M., Eds.; Springer: Cham, Switzerland, 2018; pp. 89–107.
15. Iglesias, O.; Markovic, S.; Bagherzadeh, M.; Singh, J.J. Co-creation: A Key Link Between Corporate Social Responsibility, Customer Trust, and Customer Loyalty. *J. Bus. Ethics* **2020**, *163*, 151–166. [CrossRef]
16. Carroll, A.B. A three-dimensional conceptual model of corporate social performance. *Acad. Manag. Rev.* **1979**, *4*, 497–505. [CrossRef]
17. Carroll, A.B. Managing ethically with global stakeholders: A present and future challenge. *Acad. Manag. Perspect.* **2004**, *18*, 114–120. [CrossRef]
18. Carroll, A.B.; Buchholtz, A. *Business and Society: Ethics, Sustainability and Stakeholder Management*, 7th ed.; Cengage Learning: Mason, OH, USA, 2009.
19. Bridoux, F.; Stoelhorst, J.W. Stakeholder relationships and social welfare: A behavioral theory of contributions to joint value creation. *Acad. Manag. Rev.* **2016**, *41*, 229–251. [CrossRef]
20. Freeman, R.E. *Strategic Management: A Stakeholder Approach*, 1st ed.; Pitman Publishing: Boston, MA, USA, 1984.
21. Friedman, M. A Friedman doctrine: The social responsibility of business is to increase its profits. *The New York Times Magazine*. 13 September 1970. Available online: https://nyti.ms/1LSi5ZD (accessed on 2 October 2020).
22. Mitchell, R.K.; Agle, B.R.; Wood, D.J. Toward a theory of stakeholder identification and salience: Defining the principle of who and what really counts. *Acad. Manag. Rev.* **1997**, *22*, 853–886. [CrossRef]
23. Miles, S. Stakeholders: Essentially contested or just confused? *J. Bus. Ethics* **2012**, *108*, 285–298. [CrossRef]

24. Lépineux, F. Stakeholder theory, society and social cohesion. *Corp. Gov.* **2005**, *5*, 99–110. [CrossRef]
25. Brugann, J.; Prahalad, C.K. Cocreating Businesses' New Social Compact. *Harv. Bus. Rev.* **2007**, *85*, 80–90. [PubMed]
26. Edward Freeman, R. Stakeholder Management. Available online: https://redwardfreeman.com/stakeholder-management (accessed on 20 November 2020).
27. Richardson, J. The business model: An integrative framework for strategy. *Strateg. Chang.* **2008**, *17*, 133–144. [CrossRef]
28. Wirtz, B.W.; Pistoia, A.; Ullrich, S.; Göttel, V. Business models: Origin, development and future research perspectives. *Long Range Plann.* **2016**, *49*, 36–54. [CrossRef]
29. Zott, C.; Amit, R.; Massa, L. The business model: Recent developments and future research. *J. Manag.* **2011**, *37*, 1019–1042. [CrossRef]
30. Osterwalder, A.; Pigneur, Y. *Business Model Generation: A Handbook for Visionaries, Game Changers, and Challengers*, 1st ed.; John Wiley & Sons: Hoboken, NJ, USA, 2010.
31. Stubbs, W.; Cocklin, C. Conceptualizing a "sustainability business model". *Organ. Environ.* **2008**, *21*, 103–127. [CrossRef]
32. Lozano, R. Sustainable business models: Providing a more holistic perspective. *Bus. Strat. Environ.* **2018**, *27*, 1159–1166. [CrossRef]
33. Upward, A.; Jones, P. An ontology for strongly sustainable business models: Defining an enterprise framework compatible with natural and social science. *Organ. Environ.* **2016**, *29*, 97–123. [CrossRef]
34. Elkington, J. Enter the triple bottom line. In *The Triple Bottom Line: Does It All Add up*, 1st ed.; Henriques, A., Richardson, J., Eds.; Taylor and Francis London: Earthscan, UK, 2004; Chapter 1; pp. 1–16.
35. Freeman, R.E. Managing for stakeholders: Trade-offs or value creation. *J. Bus. Ethics* **2010**, *96*, 7–9. [CrossRef]
36. Grigorescu, A.; Maer-Matei, M.M.; Mocanu, C.; Zamfir, A.M. Key Drivers and Skills Needed for Innovative Companies Focused on Sustainability. *Sustainability* **2020**, *12*, 102. [CrossRef]
37. Osterwalder, A.; Pigneur, Y.; Tucci, C.L. Clarifying business models: Origins, present, and future of the concept. *Commun. Assoc. Inf. Syst.* **2005**, *15*, 1–40. [CrossRef]
38. Kaplan, R.; Norton, D. The Balanced Scorecard—Measures that Drive Performance. *Harv. Bus. Rev.* **1992**, *70*, 71–79.
39. Kaplan, R.S.; Norton, D.P. Using the balanced scorecard as a strategic management system. *Harv. Bus. Rev.* **1996**, *74*, 75–85.
40. Lüdeke-Freund, F.; Freudenreich, B.; Saviuc, I.; Schaltegger, S.; Stock, M. Sustainability-Oriented Business Model Assessment—A Conceptual Foundation. In *Analytics, Innovation and Excellence-Driven Enterprise Sustainability*, 1st ed.; Edgeman, R., Carayannis, E., Sindakis, S., Eds.; Palgrave: Houndmills, UK, 2017; Chapter 7; pp. 169–206.
41. Hansen, E.G.; Schaltegger, S. The Sustainability Balanced Scorecard: A Systematic Review of Architectures. *J. Bus. Ethics* **2014**, *133*, 193–221. [CrossRef]
42. Figge, F.; Hahn, T.; Schaltegger, S.; Wagner, M. The Sustainability Balanced Scorecard—Linking Sustainability management to Business Strategy. *Bus. Strat. Environ.* **2002**, *11*, 269–284. [CrossRef]
43. Jones, P.; Comfort, D.; Hillier, D. Managing Materiality: A Preliminary Examination of the Adoption of the New GRI G4 Guidelines on Materiality within the Business Community. *J. Public Aff.* **2016**, *16*, 222–230. [CrossRef]
44. Starik, M.; Marcus, A. Introduction to the Special Research Forum on the Management of Organizations in the Natural Environment: A Field Emerging from Multiple Paths, With Many Challenges Ahead. *Acad. Manag. J.* **2017**, *43*, 539–547. [CrossRef]
45. Phillips, R.; Reichart, J. The Environment as a Stakeholder? A Fairness-Based Approach. *J. Bus. Ethics* **2000**, *23*, 185–197. [CrossRef]
46. ACCA Global. 2013 ACCA, Flora & Fauna International and KPMG LLP. Identifying Natural Capital Risk and Materiality. Available online: https://www.accaglobal.com/gb/en/technical-activities/technical-resources-search/2014/january/identifying-natural-capital-risk-and-materiality.html (accessed on 22 November 2020).
47. Alonso-Almeida, M.M.; Llach, J.; Marimon, F. A Closer Look at the 'Global Reporting Initiative' Sustainability Reporting as a Tool to Implement Environmental and Social Policies: A Worldwide Sector Analysis. *Corp. Soc. Responsib. Environ. Manag.* **2014**, *21*, 318–335. [CrossRef]
48. Dennis, P.; Connole, H.; Kraut, M. The efficacy of voluntary disclosure: A study of water disclosure by mining companies using the global reporting initiative framework. *J. Leg. Ethical Regul. Issues* **2015**, *18*, 87–106.
49. Marimon, F.; Alonso-Almeida, M.D.; Rodríguez, M.D.; Alejandro, K.A.C. The worldwide diffusion of the global reporting initiative: What is the point? *J. Clean. Prod.* **2012**, *33*, 132–144. [CrossRef]
50. Toppinen, A.; Li, N.; Tuppura, A.; Xiong, Y. Corporate Responsibility and Strategic Groups in the Forest-based Industry: Exploratory Analysis based on the Global Reporting Initiative (GRI) Framework. *Corp. Soc. Responsib. Environ. Manag.* **2012**, *19*, 191–205. [CrossRef]
51. Calabrese, A.; Costa, R.; Ghiron, N.L.; Menichini, T. Materiality analysis in sustainability reporting: A tool for directing corporate sustainability towards emerging economic, environmental and social opportunities. *Technol. Econ. Dev. Econ.* **2019**, *25*, 1016–1038. [CrossRef]
52. Ouvrard, S.; Jasimuddin, S.M.; Spiga, A. Does Sustainability Push to Reshape Business Models? Evidence from the European Wine Industry. *Sustainability* **2020**, *12*, 2561. [CrossRef]
53. Benson-Rea, M.; Brodie, R.J.; Sima, H. The plurality of co-existing business models: Investigating the complexity of value drivers. *Ind. Mark. Manag.* **2013**, *42*, 717–729. [CrossRef]
54. Christ, K.L.; Burritt, R.L. Critical environmental concerns in wine production: An integrative review. *J. Clean. Prod.* **2013**, *53*, 232–242. [CrossRef]
55. Olaru, O.; Galbeaza, M.A.; Bănacu, C.S. Assessing the Sustainability of the Wine Industry in Terms of Investment. *Procedia Econ. Financ.* **2014**, *15*, 552–559. [CrossRef]

56. Forbes, S.L.; Cohen, D.A.; Cullen, R.; Wratten, S.D.; Fountain, J. Consumer attitudes regarding environmentally sustainable wine: An exploratory study of the New Zealand marketplace. *J. Clean. Prod.* **2009**, *17*, 1195–1199. [CrossRef]
57. Dania, W.A.P.; Xing, K.; Amer, Y. Collaboration behavioural factors for sustainable agri-food supply chains: A systematic review. *J. Clean. Prod.* **2018**, *186*, 851–864. [CrossRef]
58. Yin, R.K. Discovering the future of the case study. Method in evaluation research. *Eval. Pract.* **1994**, *15*, 283–290. [CrossRef]
59. Edmondson, A.C.; McManus, S.E. Methodological fit in management field research. *Acad. Manag. Rev.* **2007**, *32*, 1155–1179. [CrossRef]
60. Eisenhardt, K.M.; Graebner, M.E. Theory building from cases: Opportunities and challenges. *Acad. Manag. J.* **2007**, *50*, 25–32. Available online: http://www.jstor.org/stable/20159839 (accessed on 5 January 2021). [CrossRef]
61. Bolis, I.; Brunoro, C.M.; Sznelwar, L.I. Work for sustainability: Case studies of Brazilian companies. *Appl. Ergon.* **2016**, *57*, 72–79. [CrossRef] [PubMed]
62. Hartley, J.F. Case studies in organizational research. In *Qualitative Methods in Organizational Research*, 1st ed.; Cassel, C., Symon, G., Eds.; Sage Publications: London, UK, 1995; pp. 208–229.
63. Viña Concha y Toro 2020a. Sustainability Report 2017. Available online: https://conchaytoro.com/content/uploads/2021/01/REPORTE-VCT-2017_ENG.pdf (accessed on 26 October 2020).
64. Viña Concha y Toro 2020b. Sustainability Report 2018. Available online: https://conchaytoro.com/content/uploads/2021/01/REPORTE-VCT-2018_ENG.pdf (accessed on 26 October 2020).
65. Viña Concha y Toro 2020c. Sustainability Report 2019. Available online: https://conchaytoro.com/content/uploads/2021/01/REPORTE-VCT-2019_ENG.pdf (accessed on 26 October 2020).
66. Beck, A.C.; Campbell, D.; Shrives, P.J. Content analysis in environmental reporting research: Enrichment and rehearsal of the method in a British–German context. *Br. Account. Rev.* **2010**, *42*, 207–222. [CrossRef]
67. Merkl-Davies, D.M.; Brennan, N.M. Discretionary disclosure strategies in corporate narratives: Incremental information or impression management? *J. Account. Lit.* **2007**, *27*, 116–196. Available online: http://hdl.handle.net/10197/2907 (accessed on 5 January 2021).
68. Guidry, R.P.; Patten, D.M. Market reactions to the first-time issuance of corporate sustainability reports. *Sustain. Account. Manag. Policy J.* **2010**, *1*, 33–50. [CrossRef]
69. Impact Management Project. Statement of Intent to Work Together towards Comprehensive Corporate Reporting. 2020. Available online: https://impactmanagementproject.com/structured-network/statement-of-intent-to-work-together-towards-comprehensive-corporate-reporting/ (accessed on 11 December 2020).

Article

Development of a New Business Model to Measure Organizational and Project-Level SDG Impact—Case Study of a Water Utility Company

Paul Mansell [1,2,*], Simon P. Philbin [2] and Tim Broyd [1]

[1] Bartlett School of Construction and Project Management, University College London, London WC1E 7HB, UK; tim.broyd@ucl.ac.uk

[2] Nathu Puri Institute for Engineering & Enterprise, London South Bank University, London SE1 0AA, UK; philbins@lsbu.ac.uk

* Correspondence: paulrmansell@gmail.com

Received: 11 June 2020; Accepted: 5 August 2020; Published: 10 August 2020

Abstract: Achievement of the United Nations' 2030 Global Goals for Sustainability is of paramount importance. However, for engineers and project managers to take meaningful action, they need the practical tools, processes and leadership to turn grand rhetoric into viable engineering solutions. Linking infrastructure project sustainability performance to sustainable development goals (SDG) targets is problematic. This article builds on the previous development of an innovative infrastructure business model, called the "Infrastructure SDG Impact-Value Chain" (IVC) to link local-level project delivery with global-level SDG impacts. It uses a case study of a water utility company to demonstrate how the IVC business model can integrate the "triple bottom line" to ensure the balanced definition of success across economic, environmental and social thematic areas. The results led to a proposed methodology for business leaders to align stakeholders on a common definition of project success during the design phase. The study includes the selection of longer-term outcomes and strategic SDG impacts, which, it is suggested, are improved definitions of project success. Although the findings that are from a single case study cannot automatically be extended to the entire water industry, the study's methodology has potential to be used to evaluate multiple projects across different sectors. The practical application is significant since it offers the flexibility to be used at both project and portfolio levels, thereby linking tactical delivery to organisational SDG impacts and leading to improved investment decisions with increased likelihood of success in achieving the SDG 2030 targets.

Keywords: sustainable development goals (SDGs); sustainability; sustainable development; project success; infrastructure project; strategy; public management

---

1. Introduction

The construction industry has a major role in achieving a measurable impact against the sustainable development goals (SDG) 2030 targets. The estimated USD $94 trillion [1] of investment in infrastructure projects that is required globally by 2040 represents a massive opportunity to stimulate economic prosperity, reduce poverty and raise standards in health, education and gender equality. However, the linking of infrastructure project success to SDG targets is problematic as a recent Institution of Civil Engineers' survey [2,3] demonstrated: while the appetite for SDG reporting at project level is very strong (87%), especially among millennials, only a third of the 325 survey respondents assessed current tools as "fit for purpose". The research study identified four critical success factors (CSF) for measuring projects' SDG impacts:

- CSF #1: strategic success definition. Clear understanding of project success: is it about time, cost and scope (doing the projects right) or is it about outcomes and strategic impacts (doing the right projects) or a balance of both?
- CSF #2: holistic performance measurement tools. The need for tools that could measure traditional outputs of time, cost and scope as well as more opaque successes, such as outcomes, benefits and impacts.
- CSF #3: aligned business priorities. Balancing competing business priorities, which were perceived to weight "profit" too heavily against "people" and "planet", otherwise known as the "triple bottom line" [4–6].
- CSF #4: strong leadership. The need for leaders who can galvanize and motivate their teams, capturing their "heads and hearts" to drive forward changed behaviours.

The shortcomings of not having the four CSFs in place, which was the main finding from the survey, represents both a theoretical knowledge gap and, for the practitioner, it results in weaker investment decisions since SDG lessons are not being learned from project delivery successes and failures. The problem is complex and multifaceted in nature at both the project and organisational levels. At its core, the most important issue is to understand what defines project success. Too often this has been done by measuring the project management processes of delivering a project to time, cost and scope (and quality), otherwise known as the "iron triangle". However, for linkage to the SDGs, there needs to be a broadening of the success definition to become more holistic. In short, it needs a new business model. To do so requires a refresh of underpinning theories, specifically in regard to sustainable development.

Before the paper addresses the specific nature of the SDGs and their potential to be used to improve project success definition, on a broader canvas than just "time-cost-scope", the paper will briefly review the definition of sustainability and also introduce sustainability measurement on infrastructure projects. It will discuss these areas in the following three subsections: the definition of infrastructure (which is the sector that the case study is situated in), the concept of sustainability and sustainable development, and definition and measurement of sustainable infrastructure at organisational and project levels.

1.1. *Defining 'Infrastructure' (the Relevant Sector for the Case Study)*

The Global Commission on the Economy and Climate defined infrastructure as: "structures and facilities that underpin power and other energy systems (including upstream infrastructure, such as the fuel production sector), transport, telecommunications, water and waste management. It includes investments in systems that improve resource efficiency and demand-side management, such as energy and water efficiency measures. Infrastructure includes both traditional types of infrastructure (including energy to public transport, buildings, water supply and sanitation) and, critically, also natural infrastructure (such as forest landscapes, wetlands and watershed protection)" [7,8].

1.2. *Sustainability and Sustainable Development*

In order to understand the SDGs, it is first necessary to explore the concepts of sustainability and sustainable development that jointly inform much of the nomenclature surrounding SDGs. Research into the definition of sustainability has indicated [9] that there are in excess of 50 separate definitions of sustainability. This highlights that there is a lack of agreement on the practical and theoretical derivation. As an example, Sverdrup and Rosen [10] suggest that sustainability and sustainable development implies the longer term harnessing of the ecosystem to a point at which the resource-capital base, framework or application of the ecosystem is not damaged or adversely changed. Conversely, Costanza and Patten [11] believe that the essence of sustainability is that it provides a litmus test to indicate whether a system survives or perishes. It can thus be shown that sustainability has become mired in value-laden language and is often vague in concept [12], which can cause diffusion of interpretation and confusion in practice [13]. Potentially, this is the reason that Glavic and Lukman [14] suggested that defining sustainable development in a practical way can be somewhat uncertain since there are several interpretations that can be deployed.

Over the past 50 years, the phraseology and understanding of "sustainable development" [15,16] has become an increasingly central theme of nation states and their citizens. Unlike sustainability, the definition of sustainable development at least has a generally agreed definition from the report of the Brundtland Commission [17]. According to the Commission, it can be defined as "development that meets the needs of the present without compromising the ability of future generations to meet their own needs" [17]. Building on this definition, sustainability and sustainable development embody a connectivity with ecological (i.e., planet) and social (i.e., people) as well as economic (i.e., profit) systems. Today, the planetary boundaries provide a global litmus test for how we are doing, using the nine boundaries within which humanity can continue to develop and thrive for generations, with the latest evidence showing that we are failing on most but most critically on three [18].

1.3. *Defining and Measuring Sustainable Infrastructure at Organisational and Project Levels*

The earlier definition of sustainable infrastructure by Ainger and Fenner [19] was recently developed further by the Inter-American Development Bank (IDB) Group as "infrastructure projects that are planned, designed, constructed, operated, and decommissioned in a manner to ensure economic and financial, social, environmental (including climate resilience), and institutional sustainability over the entire life cycle of the project" [20]. The focus of their investigation was on the detailed analysis of existing sustainability reporting methods across two of the hierarchy levels, i.e., at the project and organisational levels. While there are literally hundreds of sustainability methods used globally, from simple spreadsheet-based approaches to enterprise-wide, cloud-based systems, there are few comparisons of these tools with methods for measuring SDG impacts. Following on from the IDB research, a recent paper by Mansell et al. [21] partially closes this gap by completing a deep and broad analysis of relevant measuring tools. Their work established a golden thread from CEEQUAL (which was compared with other global project measuring tools) with links to the Global Reporting Initiative's (GRI) global standard for organisational sustainability measurement. Importantly, their research shows both can be linked to SDGs, although both are at a nascent stage of doing so. The research was conducted with the collaboration of both GRI and the Building Research Establishment (BRE), which is UK's leading centre of building science. BRE, as the owners of CEEQUAL, gave full access to their systems and standards to enable completion of the detailed text and process analysis of both standards in comparison to the SDG targets and indicators. A summary of some of the leading sustainability reporting frameworks from this study [21], at organisational and project levels, is shown in Table 1 with a brief analysis of their explicit or implicit alignment with SDG measurement. It does not purport to provide a full in-depth comparison or discussion of the relative merits, which can be found in the Mansell et al. paper [21].

Table 1. Summary of some of the leading infrastructure sustainability reporting tools/methods at organisational and project levels; for full discussion on these tools and in-depth analysis of CEEQUAL and the Global Reporting Initiative (GRI), see further research by Mansell et al. [21].

Tools and Methods	Relevance for the Case Study
1. Organisational level tools and methods. Global Reporting Initiative [22], UN Global Compact [23], Carbon Disclosure Project, GHG Protocol [24], OECD guidelines and integrated reporting [25].	Based on analysis of the industry leading sustainability reporting frameworks [26,27], GRI was shown to be the most frequently used by leading companies. Indeed, of the world's largest 250 corporations, 92% report on their sustainability performance and 74% of these use GRI's standards to do so, with 23, 00 corporate sustainability reports currently in the GRI database [22,28]. For example, it was used by 6671 organisations in 2017 [22] and 75% of Fortune 250 companies across 91 countries. Whilst the UN Global Compact has the "SDG Compass" methodology to support organisations to measure SDG impacts at subnational level, it remains at a high level and does not include any accepted standards for measurement or subnational criteria. The case study expands on the challenge of trying to use the national level targets at organisational and project levels. The GRI has also tried to leverage the widely accepted framework [21] to explicitly measure SDGs but, to date, this has proved problematic since the national level measurement framework is too complex [21], with its 169 targets and 232 indicators (discussed in a later section).
2. Organisational and project level tools and methods. Thirteen sustainability assessment methods were examined, including the following: CEEQUAL (UK & Ireland Projects/International Projects) BREEAM [29], Halstar [30]; SPeAR [31], ASPIRE [32], ISO14001 [33], OHSAS 45001 [33], Jacobs Value [34], LEED [35], ENVISION rating system by ISI and Harvard University [36], IS rating scheme by the Infrastructure Sustainability Council of Australia [37], infrastructure voluntary evaluation sustainability tool (INVEST) [38], SuRe® Standard for Sustainable and Resilient Infrastructure [39], sustainable transportation appraisal rating system framework (STARS) [40], IFC Performance Standards on Environmental and Social Sustainability, and World Bank Environmental and Social Framework.	The project-level sustainability frameworks were assessed against their ability to measure SDGs. Most of these were developed before the SDGs were agreed at the UN by the 193 states in 2015 and thus have no formal linkage to SDG measurement. Some, such as CEEQUAL, have started to link to both SDGs and to the GRI to establish a golden thread from project level to organisational level to national-global levels [21]. However, although this research has confirmed there is the potential for the "golden thread" from project to global goals, there is no evidence yet found of projects and organisations having achieved this requirement. Therefore, this confirms the knowledge gap and explains why the case study in this paper is important to commence the research into how leading companies have addressed this matter (the choice of Anglian Water was motivated by their award of the UK's national prize in 2017 as "Sustainability Company of the Year").

1.4. Sustainable Development Goals

The United Nations' "Transforming Our World" report [41] was adopted by 193 states at the United Nations General Assembly. This has provided a globally agreed sustainable development framework consisting of 17 goals (as shown in Figure 1) and 169 targets to be achieved by 2030. However, progress towards the 2030 targets is perilously slow, especially for the most disadvantaged and marginalised groups [42]. While there have been some significant advances since the Rio Summit in 1992, the "+20" in 2012, and the Kyoto Protocol, such as the transformational technologies for battery-powered cars and renewable energy, even a rise of 1.5 °C now appears to be inevitable [43]. This temperature rise would potentially wipe out almost all of the world's coral with hundreds of millions of people potentially killed from the effects of drought and coastal flooding, while the threat of starvation will likely trigger unprecedented mass migration [43].

Figure 1. The United Nations 17 Sustainable Development Goals [41] (full details can be accessed at https://sustainabledevelopment.un.org/). (Usage of graphic agreed by UN).

The delivery targets are understandably ambitious and needed a reporting framework that would drive meaningful and verifiable progress towards the 2030 targets. In 2017, the UN's Inter-agency Expert Group on Targets and Indicators for Sustainable Development designed a mechanism that linked goals, targets and indicators across the geographic and governance boundaries at national, regional and global levels [44]. Within this framework, shown in Figure 2, the Expert Group designed thematic areas that could also be used at the subnational level but, because the targets and indicators were originally designed to be used at global, regional and national level, they had reduced applicability at organisational or project levels. Simply stated, "one size does not fit all". This provides a significant challenge because most of the investment needed (USD $94 trillion) to respond to the global goals [1] is delivered through the business sector, typically through infrastructure projects, which contribute to the systems and services that can positively impact health, wealth and inequalities.

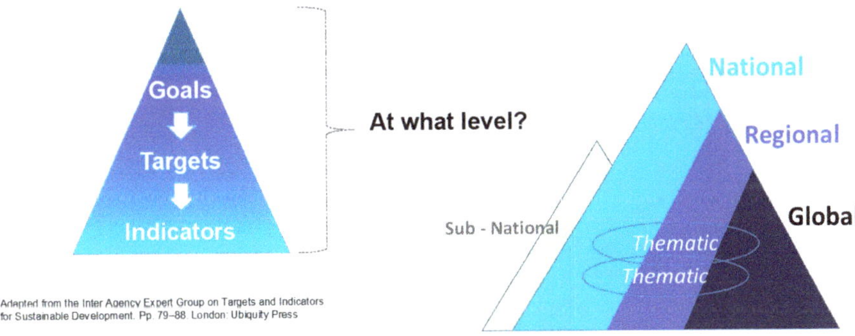

Figure 2. The sustainable development goals (SDG) Targets and Indicators' framework designed by the UN IAEG-SDGs [44].

As stated earlier, the SDGs consist of 17 major goals and 169 concrete targets and, because some of the targets are not expressed as concrete numbers, the UN also developed a framework of 232 indicators for monitoring and reviewing the targets. Research into the use of the SDG framework [21,45] on infrastructure projects has identified that the targets (N = 169) and indicators (N = 232) are too numerous and complicated and therefore, unfortunately, they are rarely used by engineering practitioners. The research concluded that a new way was needed to reduce the scientific and statistical complexity of the SDG measurement framework. The starting point for this approach was to evaluate their usability and applicability at the project level on a sector-by-sector basis. For example, in the infrastructure sector, recent analysis [46] indicates that 81% of the SDG targets are influenced by infrastructure investment projects. However, "influence" is a comparatively weak word without specifying "attribution" (i.e., directly impacting with verifiable evidence) or "contribution" (i.e., linkage presumed but without evidence) and, therefore, despite the positive conclusion from the UNOPS's analysis [46], further research is needed to identify which of the SDG targets can be used at project level. This provides a fifth CSF:

Additional critical success factor for measuring projects' SDG impacts (#5): prioritisation of (a limited) number of SDG targets relevant to the infrastructure project.

The problem of identifying suitable SDG measurement is compounded at the indicator level, where a further 232 measurement metrics reside. For example, the UK's Office for National Statistics (ONS) online portal, responsible for reporting UK's progress against global SDG indicator measurement, shows that, in April 2019, they only had data for 173 of the 232 indicators, with 69 being without data [47]. The ONS's challenge of collating reporting evidence for the 232 indicators was further corroborated by recent analysis [45] of the viability of using each of 232 indicators for infrastructure project-level measurement of success. The analysis, based on inductive reasoning using the project success framework proposed by Morris [48] and Cooke-Davies [49] and then analysed against the cost-benefit measurement framework from the HMT Green & Orange Books [50] and the World Bank monitoring, reporting, evaluation and learning methodology [51], highlighted there were only a small number of indicators (N = 28; 12%) relevant to engineering projects. Of these, only 8% (N = 20) have close alignment with the engineering projects and 4% (N = 8) have marginal relevance, as shown in Figure 3. This analysis highlighted a "gap" of not having suitable indicators below the SDG target level that could be used on infrastructure projects.

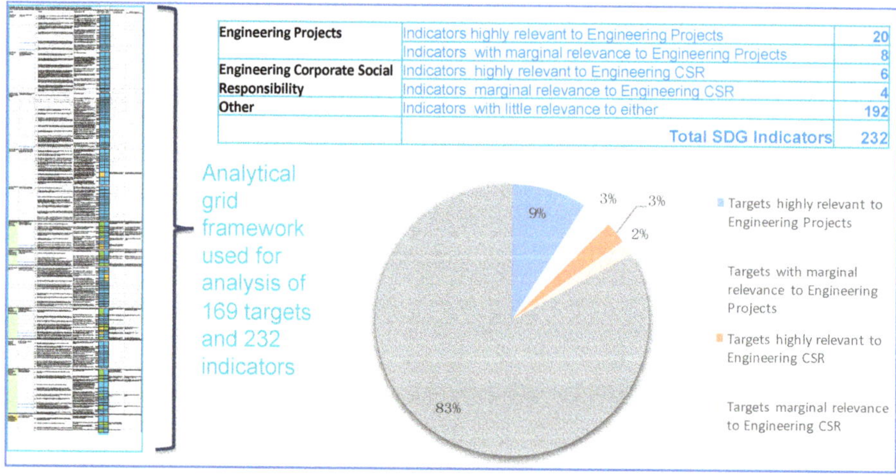

Figure 3. Analysis of the SDG Targets and Indicators' measurability.

Additional critical success factor for measuring projects' SDG impacts (#6): selection of (a limited) set of specific infrastructure indicators (not SDG indicators) relevant for infrastructure projects.

1.5. Project Success

Before sharing the new process model, it is important to reflect on the different ways of defining project success, particularly since its relevance is linked to two of the original critical success factors: critical success factor #1 (strategic success definition) and critical success factor #3 (aligned business priorities). While project success is a heavily researched field of study within the field of project management (see for example the work of [52,53] the quantitative analysis of success criteria, and their alignment to outputs or outcomes, is less evident. For example, Thiry [52] highlights that "too many critical success factors are related to inputs and management processes and not enough on outcomes". This is further supported by those [48,49] who identify two primary levels of success criteria: project management success (was the project done right?) and, secondly, project success (was the right project done?). To explain the difference, it is helpful to go back to basics—that projects are temporary organisations that have a well-recognised development process, referred to as the project life cycle [48]. To achieve its "ends" (post project), the project management team harnesses the "ways" of tools and techniques, and employs practices, processes and procedures by "means" of a group of skilled individuals. Together the ends, ways and means form a distinct body of knowledge, such as the APM's and PMI's body of knowledge. There is, however, a fundamental problem that, as a discipline, project management too often defines success by the best use of these practices instead of what its impact is on producing outcomes of real value [48]. This is important to resolve because of the huge investment across all projects to effect successful change, especially when related to strategic SDG impacts. The two fundamental parts of defining project success are shown in Figure 4. The first question is focused on the delivery phases and is tactical in nature, while the second seeks to define the longer-term outcomes and impacts, which are more strategic in orientation.

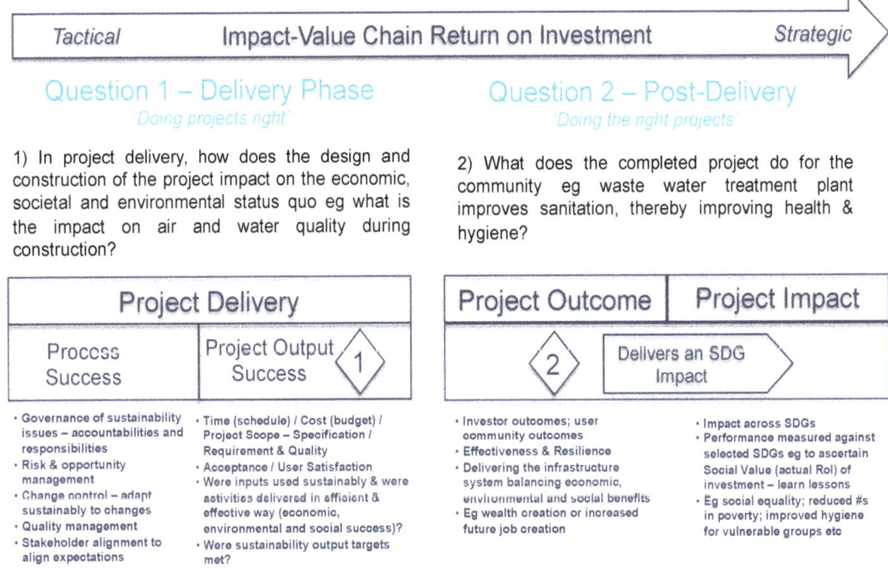

Figure 4. Framework for sustainability and project success reporting. The two core sustainable development questions at project level.

1.6. Infrastructure SDG Impact-Value Chain (IVC) Process Model

Having defined the different ways of classifying project success, a new SDG business model was developed for the infrastructure sector [21]. It provides the "lens", called the SDG infrastructure impact-value chain (IVC), to analyse whether there is evidence of a "golden thread" between best practice sustainability reporting frameworks at project and organisational levels and those at strategic-level SDG impacts.

The IVC model (see Figure 5) is based on four underpinning theoretical models including: (1) the Theory of Change [54,55], (2) creating shared value [56,57], (3) infrastructure systems approach [58–61] and (4) the triple bottom line [4–6]. The last of these, the TBL, provided the link to SDGs through a more holistic "systems approach" to address infrastructure sustainability in the SDG context. The IVC provides a new holistic method to potentially improve sustainability on projects and programmes by guiding decision makers in their investment choices through confidence that they link to specific SDG targets.

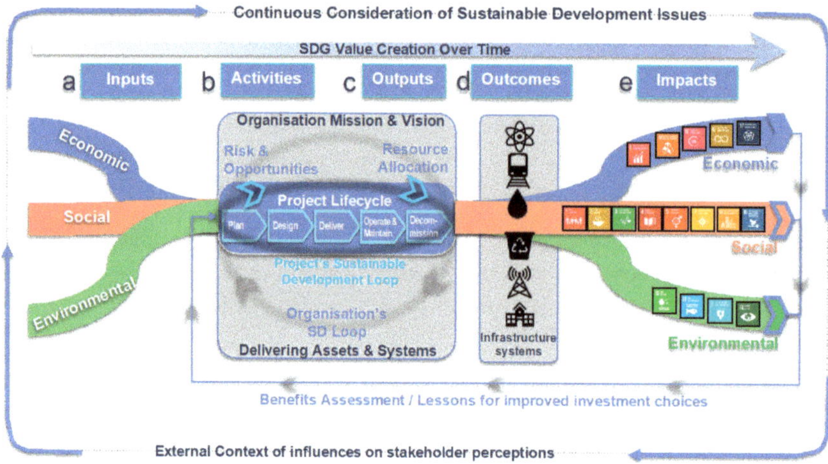

Figure 5. The infrastructure SDG transformation process model—the impact-value chain (IVC). Adapted from ICAS/IIRC's "The Sustainable Development Goals, integrated thinking and the integrated report" [62].

In practice, the golden thread (the TBL thematics of economic, social and environmental), shown in Figure 5, can be used to map the TBL against the five stages of the IVC as shown in Table 2 (with columns a–e also represented in Figure 5). The examples provided indicate that there are clear "Theory of Change" [54,55] patterns that build through the iterative stages and this can be linked directly to project- and organisational-level understanding of sustainability reporting.

Table 2. IVC table illustrating golden thread mapping of the TBL with the five stages of the IVC.

	(a) Input	(b) Activity	(c) Output	(d) Outcome	(e) Impact
Economy	Finance/investment, insurance, risk contingency allocations, WLC analysis, stable government and noncorrupt financial context.	Job creation; income; wages; source, move and assemble materials; build iteratively through defined activities, such as early earthworks, and local and wider supply chain activity	Project completion to time/cost/scope—bridge, building, road, etc.; income; profit; taxes from in-project business and net present value provides strong RoI against whole life costs.	Economic growth enabled by completed assets as a system, more resilience, wealth creation, ownership, increased future investment and additional job creation.	SDGs 8, 9, 10 and 12.
Social	People, social networks, cultural and technical knowledge, and listening and working with stakeholders.	Collaborative innovation, health and wellbeing, stakeholder engagement, skills and learning, working conditions, production activity and user engagement.	Asset's social utility, meeting stakeholders' objectives, individual and group learning, and reinforced community stakeholder groups.	Infrastructure enabled change across health, education, etc., e.g., reduced mortality; gender equality; social equity; justice and post-project knowledge sharing.	SDGs 1, 2, 3, 4, 5, 7 and 11.

Table 2. *Cont.*

	(a) Input	(b) Activity	(c) Output	(d) Outcome	(e) Impact
Environment	Raw materials, land take, water, light, clean air, energy, planned land use and ecology ecosystem valuation assessment.	GHG emissions; pollution; noise and air quality and works' effects pre and during production, e.g., waste management, nitrogen, carbon dioxide and acidification levels.	Managed effects on completion of asset; replanted trees, etc.; improved local area; no net loss on eco system footprint and short-term environmental targets met.	Restored/improved biodiversity and natural balance, e.g., increased long-term positive effect on environment through improved sustainability.	SDGs 6, 13, 14 and 15.

The data in Table 2 provide the conceptual basis for proposing that there is a golden thread that links tactical success during delivery to the strategic success embodied in the post-project outcomes and SDG strategic impacts.

The next section uses a case study of a UK water utility company, Anglian Water, to demonstrate how the IVC process model can integrate the "triple bottom line" [4–6] to ensure balanced definition of success across economic, environmental and social thematics. The emphasis is switched from "doing projects right" to "doing the right projects"; both are important, but the latter is critical. This is an explicit part of the IVC model, ensuring that short-term project success measures are balanced with post-project longer term outcomes and SDG strategic impact, which many [48,49] have suggested are improved definitions of project success.

2. Methods

The preceding literature review provided insights into the specific research problem of infrastructure project SDG measurement. The review included several themes (infrastructure, sustainability and sustainable infrastructure) as well as reviewing pre-2015 (when the SDGs were agreed at the UN) sustainability measurement methodologies and tools. The learning derived from the literature review illustrated the knowledge gap that exists when using previous sustainability tools, which were not designed for the SDGs, indicating that their use on SDG measurement is uncertain. The case study enables an opportunity to assess an approach by a leading UK water utility company to close the gap. This approach is consistent with what May [63] identified as the fact "that literature should support the researcher in designing and planning the frameworks for the research". In this way, the literature review enabled the choice of the methodology.

2.1. Using the Realist Evaluation Methodology to Structure the Research

The research study adopts the critical realism perspective of ideological philosophers, such as Bhaskar [64], to inform the choice of the realist evaluation approach, primarily because of its practical utility and its widespread use in social science research into the impacts of programmes [65]. It also provides a way to develop theory-led investigations, which is what this research seeks to do on SDG measurement. The adoption of the realist evaluation's context-mechanism-outcome (C-M-O) configuration [66,67] is widely used across clinical research and increasingly across social sciences [65]. Indeed, Pawson and Tilley specifically recommend the C-M-O strategy so that "programme theories can be tested for the purposes of refining them" [66] (p. 12). In this regard, the investigation is not about what works but asks instead "what works for whom in what circumstances and in what respects, how?" [66] (p. 2). Therefore, this research approach provides a strong framework for analysing engineers' perceptions of the context of SDG measurement as well as the potential outcome on redefining investment decisions to achieve broader SDG impacts. For the purposes of this study, the definitions of C-M-O are:

- Context: the conditions in a context of action encompass "material resources and social structures, including the conventions, rules and systems of meaning in terms of which reasons are formulated" [68].

- Mechanism: the underlying entities, processes or structures that operate in particular contexts to generate outcomes of interest [69].
- Outcome: the practical effects produced by causal mechanisms being triggered in a given context [70].

2.2. Using a Case Study to Test the Transformation Process Model

The research team's method was based on using a case study investigation to test and validate the application of SDG measurement on infrastructure projects. The starting point, as shown in Figure 6, was to establish the parameters of the research, briefly outlining the SDGs and the challenge of measuring goals, targets and indicators at project level. This led to the proposed infrastructure SDG transformation process model, called the "Infrastructure SDG Impact-Value Chain" (IVC) [21], that links tactical-level project delivery with global-level strategic SDG impacts. In the process of this analysis, it identifies six areas linked to the "context-mechanism-outcome" (C-M-O) framework that are evolved from the four critical success factors (CSF) in the survey [3], each with its own underpinning question. These CSF questions are then tested against the case study of Anglian Water, a water utilities company that has developed a new business model approach and started the process of embedding SDG reporting at both organisational and project levels. Finally, the results from the case study enable an adaption of traditional business models that have typically focused too much on short-term financial business cases for their investment decisions. It shows that, by using the IVC, the new business model approach could be used at the project design phase to align stakeholders on why/when/how/what SDG targets to measure.

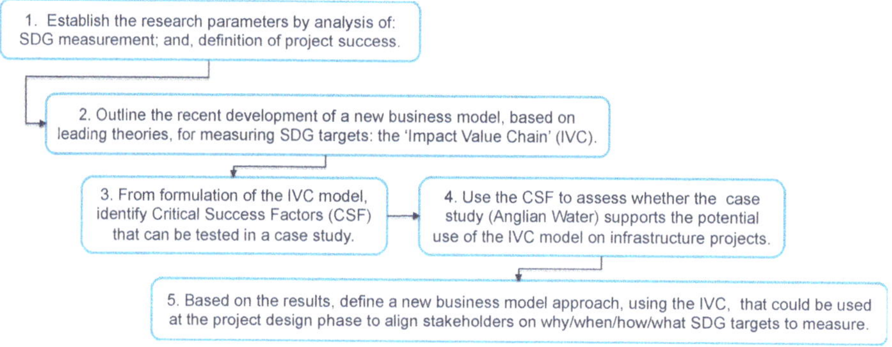

Figure 6. Research methodology employed.

As shown in steps three and four, the case study analytical approach was structured to investigate the four CSFs that were identified from the survey [3] and the two additional CSFs that have been identified from the development of the IVC model [21], as shown in the composite CSF table below (Table 3).

Table 3. Critical success factors (CSF) for embedding SDG target measurement at project level.

Category		C-M-O Critical Success Factors for SDG Measurement	Derivation
CSF enablers (context and outcome)	1	Context—strong leadership. What is the role of leadership to champion the SDG impacts across the TBL?	From engineers' survey [3]. Identified as #4 critical success factor.
	2	Outcome—clarity of IVC project success definition. Do businesses have a clear understanding of the need to separate the definition of success between "in-project" inputs/activities/outputs and "post-project" outcomes and impact?	From engineers' survey [3]. Identified as #1 critical success factor.
CSF for selection and reporting SDGs (mechanism)	3	Mechanism—step 1—prioritising SDG goals aligned to strategic vision. Do businesses have a clearly defined strategy that can guide the prioritisation of SDG goals? The "Ends, Ways, Means" model requires clarity of the "ends" prior to defining project success (in-project and post-project). See column e in Table 2.	From engineers' survey [3]. Identified as #1 and #3 critical success factors.
	4	Mechanism—step 2—select targets relevant to the project. Which SDG goals and which relevant targets are selected at project level to measure impact? Prioritisation of (a limited) number of SDG targets relevant to the infrastructure project.	From SDG analysis [45] and identified in this paper as #5 and #6 critical success factors.
	5	Mechanism—step 3—aligned business priorities/integrate the targets across the TBL. How are the project success criteria balanced across the triple bottom line and what trade-offs are made?	From engineers' survey [3]. Identified as #3 critical success factor.
	6	Mechanism—step 4—reporting and communication. Are the tools available for holistic measurement of success? What is the best way to share data on SDG progress, internally and externally?	From engineers' survey [3]. Identified as #2 critical success factor.

2.3. Central Investigation Using the C-M-O Approach

The central investigation was to test the new IVC business model against current practice using the example of one of the UK's largest water utility companies, Anglian Water. It is amongst the UK's leading sustainability and sustainable development reporting pioneers (with early use of SDG targets) and was the winner of Business in the Community's (BITC) Responsible Business of the Year Award in 2017. This recognised Anglian Water's ambitions, laid out in its "Love Every Drop" (of water) vision, which aimed to create a resilient environment that allowed sustainable growth and the ability to cope with the pressures of climate change.

The data for the case study were accessed by interviewing (1.5 h) a senior board-level member of the Anglian Water executive who, at the time, was the Director for Asset Management (DirAM). A second interview was held with the head of Anglian Water's sustainability management, as a further source of data and information. The DirAM was also the chair of the UK government's Green Construction Board's [71] Infrastructure Working Group and has been a major sponsor and champion of the sustainable development programme across Anglian Water, as well as the infrastructure sector more generally, for the past 10 years. The DirAM provided publicly available documents (i.e., as a form of secondary research) to support the in-depth insights into the company's pioneering work in sustainable development. This research was triangulated by further review and evaluation of the company's website and related documents, as well as social media, on the company's approach to sustainable development in order to verify the data's validity. Formal agreement for the review and the publication of the findings was agreed by the company in writing by DirAM and Anglian Water's Director of Brand and Communications.

3. Results and Findings

3.1. Case Study Investigation: Anglian Water—Organisational Focus on Sustainable Development

The Anglian Water approach to sustainability and the SDGs is explained in their Annual Integrated Report [72]. The report includes a description of their impact-value objectives (performance against outcomes) assessment, which correlates with the triple bottom line of the economic, social and environmental thematics. In summary, Anglian Water (AW) describe their TBL priorities as follows (Table 4).

Table 4. Anglian Water's performance against outcomes.

Anglian Water Outcomes	Objectives
1. Smart business. Innovating by exploring new ways to operate more sustainably and helping customers, business partners and employees to embrace our Love Every Drop strategy.	i. Resilient business. ii. Investing for tomorrow. iii. Fair charges, fair returns. iv. Our people: healthier, happier, safer.
2. Smart communities. Collaborating and engaging with customers, colleagues and business partners, and inspiring them to take positive steps towards achieving our vision for a sustainable future.	i. Positive impact on communities. ii. Safe, clean water. iii. Delighted customers.
3. Smart environment. Transforming behaviours by playing a leading role in reshaping how society values and uses water and reducing our combined impact on the world around us.	i. A smaller footprint. ii. Flourishing environment. iii. Supply meets demand.

These are shown below in the images from the Annual Report [72] (pp. 25, 29) (Figure 7).

Figure 7. Anglian Water alignment of purpose-outcomes and SDGs [72].

The following analysis of the case study is structured according to each of the CSF titles. The data are shown in the form of key quotes from the Director for Asset Management (DirAM) for the company, supported by data gathered from open source documents.

3.2. Context—CSF1: Strong Leadership. What Is the Role of Leadership to Champion the SDG Impacts across the TBL?

Consistent with the survey results [3], Anglian Water place a high priority on leadership to galvanise commitment to their corporate-level sustainability objectives. They achieve this through consistent and strong communications, both graphically, such as through their "Purpose Wheel" (Figure 8), and by the high-profile championing of their sustainable development approach by their board and executive.

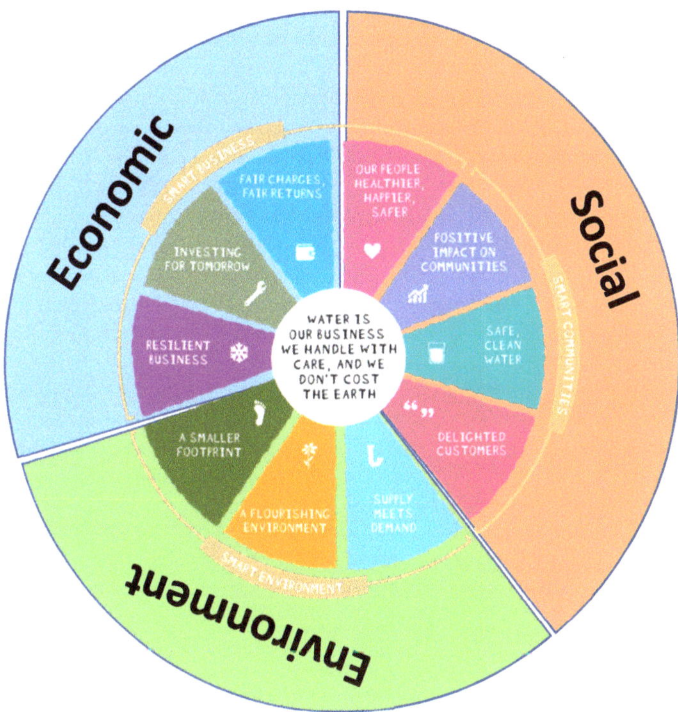

Figure 8. Anglian Water purpose wheel [72] aligned to the triple bottom line.

DirAM, a Director and Executive Board member at Anglian Water, observed (note: in future, all quotes from the interview are labelled as "DirAM" followed by the quotation): "Leadership is the most important critical success factor, both internally and externally, to align and galvanise our employees, our communities and the supply chain. It was about getting us all to be more collaborative in finding novel, innovative ways of delivering sustainable solutions . . . It is about the leaders capturing the hearts and minds of the stakeholders to champion changed behaviours to achieve big, bold strategic outcomes."

In his view, it played an important part in Anglian Water becoming a sustainable development leader across the sector. DirAM: "there are a number of reasons why we won Business in the Community's (BITC) Responsible Business of the Year Award in 2017—but a key part was that our CEO brought a very specific challenge back to the business having been inspired by a 'Seeing is Believing' visit, organised by BITC, to an area near the Olympic Park in London. The visit looked at how businesses were able to create opportunities and skills for those living in areas of high deprivation and low social mobility. The CEO's response was: 'how can we do something on a similar scale, in the region we serve, to make a real difference?'. This led to our hugely successful programme in

Wisbech and helped us develop an approach that we have subsequently used on project work in Nepal alongside Water Aid."

(Note: The Wisbech project, discussed further in Section 3.8, was a forerunner of the Lahan project in Nepal. Lahan was the first WaterAid project with significant engagement from the utilities' supply chain and became a beacon to demonstrate how such projects can be driven across Nepal and beyond.) The quote also reinforces Porter's theory of creating shared value [56,57] because, in this example, there are tangible benefits for the business to be seen to be actively "putting back" into society.

He also notes the moral values that are implicit in the choice of making sustainable development a core business priority for Anglian Water. DirAM: "a vital part of leadership is doing the right thing, just because it is the right thing to do, not because of a box-ticking exercise". DirAM expands this to state the following: "Our leadership was engaging the supply chain proactively to collaboratively change the way we thought about, and did, our business ... We wanted the approach to become part of the way we jointly became leaders in delivering our businesses successfully ... We wanted to establish meaningful change across the supply chain, and we recognised that, to do this, we had to develop long-term relationships; hence, we contracted on a five-, plus five-, plus five-year basis. This built longevity into our thinking and allowed true innovation to develop solutions to the bigger sustainable development issues across the environment—driving efficiency and effectiveness."

This was not necessarily an approach that was either quick or easy and it needed a tough commitment from the leadership; DirAM: "It is 50% belief and 50% belligerence when you start something like this; that is, holding yourself and others to account. That is what I mean by belligerence. In other words, 'seeing it through'."

The core principles of governance [73] of accountability, responsibility and transparency were also noted; DirAM: "a key part of the leadership is the ownership of the sustainable development strategy. It is also about accountability and having the resources to deliver the solution. That is why the 'Infrastructure Clients' are the single most important stakeholders in addressing sustainable development. If they 'own' and champion the solution, then the supply chain will follow ... hence, leadership and procurement are the biggest elements of the recent Green Construction Board's 'Three Years On Report—Reducing Carbon Reduces Cost' report" [71].

3.3. Outcomes—CSF 2: Clarity of IVC Project Success Definition. Do Businesses Have a Clear Understanding of the Need to Separate Definitions of Success between "In-Project" Inputs/Activities/Outputs and "Post-Project" Outcomes and Impact?

In the Anglian Water Integrated Report 2018, [72] (p. 8), the CEO says: "We are continuing to plan and to invest in protecting customers and the environment. This year saw the publication of our draft Water Resources Management Plan, which sets out how we propose to balance supply and demand in a fast-growing region over the next 25 years and to protect customers from severe water restrictions in a future drought." The Annual Report highlights that Anglian Water explicitly assesses both the short-to-medium term economic factors that their investors value as well as the longer term strategic sustainable development impacts that are more aligned to SDG targets.

DirAM explains how Anglian Water used the overall "Love Every Drop" banner campaign to balance long-term and short-term priorities: "In 2015 we refreshed our 'Love Every Drop' goals and aligned them with the Outcomes Wheel shown in the Annual Report. So, we thought long and hard about not just the goals that we created but how that fit with a set of longer-term outcomes in our region and what that would look like in terms of implementation. This was our way of meaningfully connecting the strategy with outcomes that our stakeholders recognised."

It was also noted that Anglian Water uses simple and accessible language (see CSF 6 on communications) to explain their "Purpose Wheel" and its linkage to outcomes-impacts. This aligns with the IVC model and indicates a viable way of thinking "big and long" whilst managing the activities and outputs on a short-term basis to track progress.

3.4. Mechanism—CSF 3: Prioritising SDG Goals Aligned to Strategic Vision. Do Businesses Have a Clearly Defined Strategy that Can Guide the Prioritisation of SDG Goals? The "Ends, Ways, Means" Model Requires Clarity of the "Ends" Prior to Defining Project Success (In-Project and Post-Project)

The Anglian Water approach aligns closely with the IVC model, since it also uses an "Ends, Ways, Means" logic similar to the Theory of Change concept (Figure 5), [54,55]. DirAM: "you must start with the end in mind, even if you have not got a detailed routemap to deliver at every stage of the journey. Part of the mantra is to set big audacious goals and then adopt an attitude of 'I have started so I will finish' and, by the way, you never actually finish, because the end goal is moving; it is like you achieve one peak but realise it is a false horizon, and so you continue your climb to the next summit."

As well as the ten prioritised goals, Anglian Water have also prioritised 35 targets that are most easily measured at project level, which are reproduced below (Figure 9).

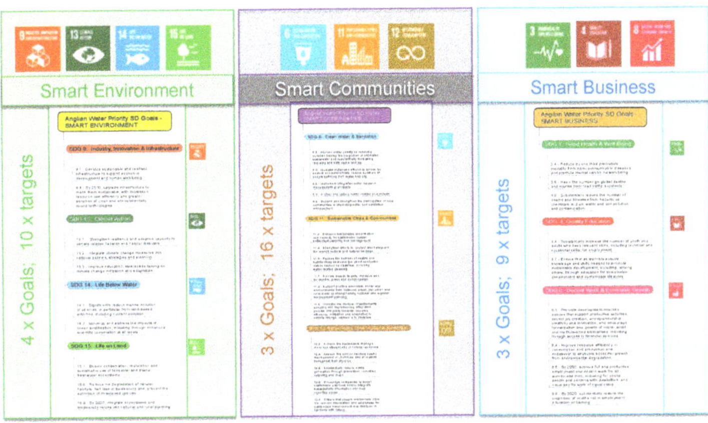

Figure 9. Anglian Water has three business priorities that are balanced across the triple bottom line (for illustration only). The specific SDG targets (N = 35) in this figure are reproduced in readable format in Table 5.

The value of having clarity of the strategic ends is noted, albeit with a caution that the identification of targets for tracking performance must not become a "box-ticking" exercise that distorts clarity of outcomes; DirAM: "if you actually begin with the end in mind of the outcome you are seeking and how you wire your DNA to achieve that, you are far more likely to achieve those outcomes, and in so doing the boxes get ticked. But if you predicate your thinking with thoughts about just filling the boxes, you have constrained yourself."

Therefore, to overcome the box-ticking mentality, DirAM explained their approach: "Anglian Water thought long and hard about its position in the region and how we contributed strategically as a major player in the region and we created the concept of "Love Every Drop" and, in essence, our own SDGs to align our strategy with local outcomes ... We used the "Love Every Drop" goals to identify ambitious aspirations, which meant that our business had to think longer term."

3.5. Mechanism—CSF 4: Select Targets Relevant to the Project. Which SDG Goals and Which Relevant Targets Are Selected at Project Level to Measure Impact? Prioritisation of (a Limited) Number of SDG Targets Relevant to the Infrastructure Project

The chart in Figure 10 illustrates the 35 targets selected by Anglian Water, which at first sight is impressive, but the interview identified that it is challenging to move beyond the rhetoric of great sounding qualitative statements. Therefore, it is important to agree and publish hard quantitative targets that the success of the organisation can be assessed against; DirAM: " ... so we nailed our colours to the mast and started reporting against those. One of them was to take 50% of the carbon out of the assets we build by 2015. It was the one that had a specific date on and a specific quantity, and I deliberately did that because I believed it and I was belligerent enough to drive it. ... That is the one that, perhaps, out of all sustainability targets and goals, Anglian Water had the greatest recognition from and probably reflects the greatest change programme that has gone on across the whole of the supply chain."

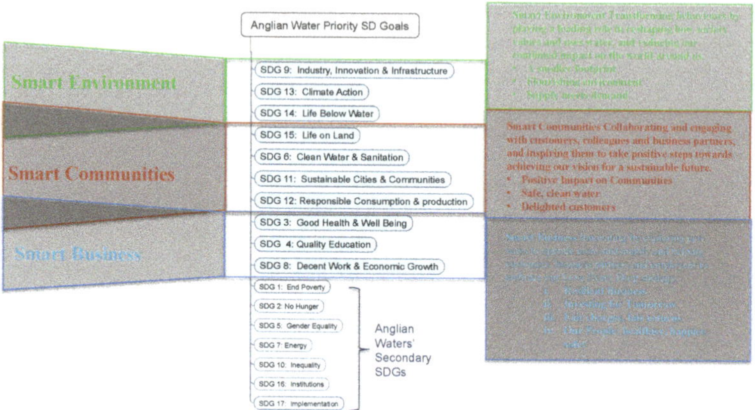

Figure 10. Anglian Water has three business priorities that are balanced across the triple bottom line.

3.6. Mechanism—CSF 5: Aligned Business Priorities/Integrate the Targets across the TBL. How Are the Project Success Criteria Balanced across the Triple Bottom Line—What Trade-Offs Are Made?

A representation of the linkage of the Anglian Water three TBL thematic outcomes [4–6], aligned to their ten prioritised SDG goals, is shown below.

In the Anglian Water integrated report of 2018 [72] (p. 9), the CEO, Peter Simpson, says: "Since becoming Responsible Business of the Year, we have been working hard to show others how sustainability makes good business sense". This quote emphasizes the Anglian Water experience that aligns with the creating shared value [56,57,74]. It implies that the TBL [75] can be balanced—a strategy that focuses on the environment and society, which can equally achieve economic success. When in harmony, real growth is delivered to the benefit of all, as shared by DirAM: "For example, our approach to 'product lifecycle management' was learned from the aeronautical and automotive industry from 2004–2005 and this meant that we looked at the whole life costs, which not only ensured we were more outcomes focused, but, by the way, improved our productivity by 3% each year, year on year, highlighting that good sustainable development also made good business sense".

3.7. Mechanism—CSF 6: Reporting and Communication. What Is the Best Way to Share Data on SDG Progress, Internally and Externally?

It has already been noted that Anglian Water had a policy of thinking long-term, explaining their sustainable development approach in accessible language and also the need to uphold strong governance principles of accountability and transparency [74]. This has led to a strong ethic of being held accountable for delivering meaningful change, including publishing their strategic objectives in quantifiable terms (such as the carbon figures noted in the paragraph above) as well as, equally importantly, the results; DirAM: "learning from the likes of Marks and Spencer's Plan A, we realised you had better publish your sustainability plans and outcome targets so that you are kept honest in the process—there is very little point nailing your colours to the mast and then not living to the high expectations ... so the message was that we must commit to do the things that matter to us. That is what gets people excited, because it really matters. We are tough on ourselves on reporting what happens, and this allows us to measure what impact we are having so that we can measure the benefit."

The theme of honesty and allowing stakeholders to hold the executive and board to account is a powerful lesson that also relates to measuring SDG impacts at project level; DirAM: "But the point about turning your ambitious goals into reality, to avoid superficial statements, is that it is all recorded—it is published annually, which is an important part of defining where you are going. Driving towards it with no 'U' turns when some tough decisions have to be made. It is obvious that you have to make loads of tough decisions rather than duck them, and then recording your progress in an open and visible way helps keep you honest in that process."

A cautionary note about communication was that the messaging should be kept simple and accessible; DirAM: "We found that our campaign and collaborative working with partners had created a different conversation with different language. Ultimately, accessible language on meaningful outcomes is what people can buy into and this is what creates the momentum of changed behaviours ... Through engagement and innovative solutions addressing the big problems, Wisbech is an example of working with the community to achieve meaningful long-term changes."

3.8. Overview Analysis of Anglian Water's Projects Set against the IVC Framework

The reference to Anglian Water's Wisbech project in the previous quote provides a holistic test against the six critical success factors and a useful way to cap the case study analysis. Launched in January 2013 as part of Anglian Water's "Wisbech 2020" vision [75], the Wisbech project was chosen as part of this case study because data on its delivery are open source on the internet. It was delivered by Anglian Water with its partners as part of their commitment to make a long-term impact on the market town of Wisbech for more than the five years that the initial project covered. Located just 40 miles from Cambridge, UK, Wisbech faced many socio-economic challenges but also had potential for significant growth and development [75]. The vision proposed a new garden town with 10,000 homes, bringing transport, education and health benefits to the town and surrounding region. By using this project as an example, Anglian Water wanted to assess whether a broad programme of social, economic and environmental change to improve the local communities' lives could be linked to the SDGs using the IVC.

The table below mirrors the formatting of the IVC table (Table 2) and has been updated with data from the Wisbech project [75]. The simple steps to achieve the Wisbech-adapted IVC included: reading and analysis of the publicly available documentation of the Wisbech project, identification of key data across the IVC framework, cross-checking across authors to assess the credibility of interpretation and sharing the final table with Anglian Water to ensure the consistency and accuracy of project data. This provides an assessment as to whether projects could have both the "in-project" successes measured as well as the "post-project" outcomes and SDG impacts as defined in the Theory of Change [54,55]. It is evident that it is easier to define quantifiable success criteria for the inputs-activity-outputs during the in-project phase because they are tangible and delivered as core delivery performance measures, such as time, cost and scope/quality. On the other hand, the outcome and impacts are typically delivered after the completion of the project and are more diffuse. Thus, the example from the Wisbech project shown below is not conclusive but gives indications that the IVC provides a useful framework to engage stakeholders on what project success looks like during and post-delivery. It should be noted that the Wisbech project is an outreach community programme inspired by HRH The Prince of Wales' "Seeing is Believing" initiative, which seeks to find ways to support marginalised communities. The SDGs therefore offer a framework to address the more diffuse outcomes and impacts that might not typically have been defined and measured using traditional project measurement approaches.

3.9. Policy Implications Derived from Analysis of Anglian Water's Use of the IVC Framework.

There are a number of policy implications, shown in Table 5, that emanate from the analysis of the Anglian Water case study. These are listed at both the organisational and project levels and involve multiple stakeholders, including clients, investors, suppliers and communities, who all benefit from the use of the derived models proposed in this paper.

Table 5. Applying Anglian Water's Wisbech project initiative to the IVC grid with mapping of the TBL with the five stages of the IVC.

	Input	Activity	Output	Outcome	Impact
Economy	Seconded a senior operational manager to Wisbech in 2013 and agreed support from other supply chain partners to become involved in the project. This allowed the cost, expertise and effort to be shared across a broad range of partners.	Worked jointly with the local Fenland District Council to develop a longer term strategy beyond their existing 2020 Vision, which was thought to be too short-term to encompass the "big, hairy, audacious" strategic goals that could achieve transformational change, building a business case for the "Garden Town" that would attract investment and large transport infrastructure improvements.	Championing apprenticeships and a training scheme with 20 trained and employed year on year. Turning the community centre from a £30, 00 per annum loss-making entity to a vital community hub, fuelling future economic success. Confirming the lease and implementing the creation of the "Jobs Fair" and the "Jobs Café" and the campaigning body for getting rail back—now in the County Transport Plan.	Bills, affordability and profits to stimulate and sustain the local economy, especially those on lower incomes (bills have only increased by 10% since 1990). Viability of the future rail and integrated transport system attracting more regional investment and raising local people's aspirations. Market town proposal, with planning for over 10, 00 new homes, providing "scale of growth" confidence.	SDGs 8, 9, 10 and 12.
Social	Started by listening in order to understand the local issues from the local community's perspective. Brought together senior leaders from "The @One Alliance", creating a collaborative multistakeholder approach. Focused on building long-term sustainable relationships with the local community.	Collaborative innovation with the local community in open and honest talks, health and wellbeing, stakeholder engagement, skills and learning, working conditions, production activity, user engagement, keeping the local community at the heart of the project plans and delivery, working with the College of West Anglia to train more mechanical and electrical engineers, designing and running new courses and providing IT support from partners to raise the aspirations of unemployed.	Providing a community centre (refurbishment of the Queen Mary Centre) that is the hub of employment opportunities; active STEM subjects engagement with schools; specifically focusing efforts on helping those not in employment, education or training; untapped, unused human resource; organised the BITC "Big Connect" event, aligning business connectors from across UK; and a second phase for the Queen Mary community Centre to include theatres and a music teaching centre.	Achieving "Business in the Community" outcomes such as regeneration; building on the "Seeing is Believing" community initiatives; understanding the value of long-term thinking; providing safe, clean and reliable water; improving the town's/region's standing as the sixth worst ranked town on the social mobility index in the UK and addressing the life expectancy that was three years less than in Cambridge.	SDGs 1, 2, 3, 4, 5, 7 and 11.
Environment	Raw materials, land take, water, light, clean air, energy, planned land use and ecology ecosystem valuation assessment.	Management plans for the flood risk, building resilience into engineering designs and using innovative modelling techniques developed by the Dutch government.	A commitment to protecting and restoring our wealth of wetland habitats and making a difference to rare and common species, be they in wet grasslands, open water, fens or mires.	Building resilience to cope with future challenges. Protecting the environment we live in. Through the Flourishing Environment Fund, helping environmental organisations deliver real benefits for nature.	SDGs 6, 13, 14 and 15.

Organisational policy implications:

- There is evidence that businesses identify value in the adoption of global SDG performance measurement at the local level. This is consistent with the theory of creating shared value [56,57,73] that identified a greater benefit to businesses than CSR being an add-on. The complexity of the global-national measurement framework makes measurement at subnational level challenging. The need for simplicity is important and examples of success, such as this case study, are helpful in galvanizing others to follow and share lessons learned. This is important for users of the models because the case study makes clear that some organisations are employing the language of SDG measurement but without a formalised methodology to do so. This makes it difficult to replicate because the ad-hoc nature of the measuring methodology used by Anglian water does not easily support cross-sector comparisons using a common framework that would have facilitated further knowledge sharing and delivery improvements.
- The SDG measurement approach can align with existing approaches to sustainability measurement. This offers efficiency of processes and systems if they can be linked. The case study gives confidence that existing reporting approaches to sustainability, such as CEEQUAL, are complementary to the proposed SDG measuring methodology. This highlights that the IVC can be adapted, such as by using language that "makes sense" to the local stakeholders and does not alienate existing project delivery teams who would not want an additional large reporting system mandated. The opportunity to align existing sustainability reporting metrics to SDG targets offers a valuable line of future research.
- There is evidence that businesses that already have a strong track record in sustainability measurement can readily adapt to the language and approach of using SDGs. Anglian Water had recently been awarded the UK's Sustainability Company of the Year, which meant that the case study interviews and review of their documentation were conducted with a highly mature organisation that had a well-developed plan for delivering clear impacts. They also had a strong leadership team to champion the trialling of the SDG measurement approach. The bigger question remains how successful the lower performing companies might be at addressing the complexities of SDG measurement. Again, this is an area for further study since that is where the majority of benefit might come from, by developing an approach that is easily replicated across the sector.
- The contextual issues, such as leadership, are a critical success factor. Strong leadership that is meaningfully engaged in championing the use of SDG measurement will be more likely to deliver tangible evidence of SDG impacts. This becomes a critical point as the strategic nature of organisational change has to be driven from the top [76]. There was recognition by the Anglian Water executive that, in reality, this meant that leaders at all levels were needed as champions, which, for SDG measurement, needed to be aligned with success stories that would make sense to the target audience written in their language and justifying "why" followed by explaining clearly "how".

Project level policy implications:

- The effective use of SDG measurement at project level needs buy-in from both internal and external stakeholders. The engagement of suppliers is critical to ensure common focus on identifying what SDG success looks like and to work collaboratively to seek innovative solutions to deliver meaningful SDG delivery success.
- There are a number of mechanistic issues that become critical to SDG measurement success. These include: prioritising relevant targets and indicators (do not select too many); seeking to understand how the few selected goals and targets can have a simple indicator framework that allows the capture of reliable evidence; and ensuring that reporting and communicating is open, honest and timely, sharing both good news and bad news. There is also a need continually to learn and evolve and so build a better framework that achieves a more balanced investment decision across the TBL of people, profit and planet [4–6,73].

4. Conclusions

The central investigation in the case study of Anglian Water was to test and validate whether the new infrastructure business model, called the "Infrastructure SDG Impact-Value Chain" (IVC), could link local-level project and organisational delivery with global-level strategic SDG impacts. The study used the "golden thread" of the TBL thematic areas (namely economic, social and environmental) to interrogate whether one of the UK's leading water utility companies, Anglian Water, was already delivering strategic sustainable development solutions that could be mapped to SDG targets. Although the research was conducted in the UK, the findings have possible broader applicability to other countries since most of the issues are neither culturally nor geographically specific. This is a valuable area of future research that could potentially engage with a number of construction firms with global footprints to compare the differences and similarities of measuring SDGs across and within different regional areas. For example, UNOPS [46] research indicates that there are many contextual global issues that affect the use and measurement of SDGs but, while noting the differences, they suggest that all issues should have a consistent framework to enable cross-cutting comparisons.

The results of the case study investigation have indicated that there is a verifiable link across the IVC of activities-inputs-outputs during the "in-project" phase, connecting to the "post-project" outcomes and SDG impacts. A number of Anglian Water's projects were mapped to this schematic (although, for brevity, only one, Wisbech, is reproduced in this article) and this gave confidence that the approach could have wider applicability. Therefore, the results led to a proposed methodology for project leaders to use as a way of strategically aligning stakeholders on a common definition of success, linking tactical "in-project" success of outputs with the more strategic outcomes and SDG impacts "post-project". The methodology would ideally be used during the design phase of the project. The emphasis is switched from "doing projects right" to "doing the right projects". It includes the selection of longer-term outcomes and strategic SDG impacts, which, it is suggested, offer improved definitions of project success.

The five proposed steps, shown in Figure 11, emanated from the six critical success factors that were used as a framework for the case study. These are proposed as a way to initiate the "right project" in the "right way" and with increased clarity of "Ends, Ways and Means".

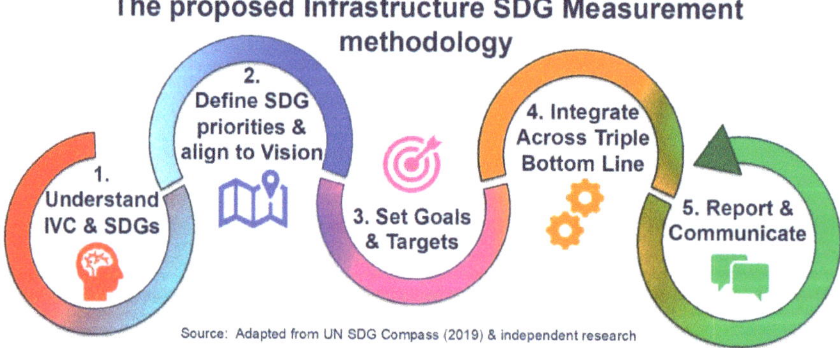

Figure 11. The proposed infrastructure SDG measurement methodology derived from the six critical success factors and the application of the impact-value chain (IVC) model to the Anglian Water case study.

Future Work

The research study has focused on a single case study in the UK and cannot automatically be extended to the entire water industry, either nationally or internationally. The methodology adopted, however, has potential to be used to evaluate multiple projects across different industry sectors. In this way, the results can thus provide insights for further research across the water industry and also potentially across other infrastructure sectors and geographical regions.

The next stage of the research is to develop the infrastructure SDG measurement methodology proposed in Figure 11 into a fully defined methodology that is adaptable to the scale of the project and also its position in the project-programme-portfolio hierarchy. Thus, the model could be tested in industrial scenarios on identified projects. The case studies will be broadened to include both developing and developed countries and will focus on a single asset type across the national economic infrastructure categories of energy, waste, water, transport and ICT. The practical application is significant since, with improved linkage of tactical delivery to strategic SDG impacts, improved investment decisions will be made, and systemic level lessons can be applied to increase the likelihood of success in achieving the SDG 2030 targets.

Author Contributions: Conceptualization, P.M.; methodology, P.M.; validation, P.M., S.P.P. and T.B.; formal analysis, P.M.; investigation, P.M.; data curation, P.M.; writing—original draft preparation, P.M.; writing—review and editing, P.M., S.P.P. and T.B.; visualization, P.M.; supervision, S.P.P. and T.B.; project administration, P.M. All authors have read and agreed to the published version of the manuscript.

Funding: This research was indirectly funded by the Nathu Puri Institute for Engineering and Enterprise, School of Engineering, London South Bank University, through the funding of the doctoral research support to Paul Mansell.

Acknowledgments: The authors would like to thank the following for their advice and support through this phase of research: Chris Newsome (former Director of Asset Management at Anglian Water Plc), Andy Brown (Head of Sustainability at Anglian Water), Ian Nicholson (Buildings Research Establishment) and the Institution of Civil Engineers (ICE).

Conflicts of Interest: The authors declare no conflict of interest. The funders had no role in the design of the study; in the collection, analyses, or interpretation of data; in the writing of the manuscript or in the decision to publish the results.

References

1. Global Infrastructure Outlook. Infrastructure Investment need in the Compact with African Countries. Available online: https://outlook.gihub.org/?utm_source=GIHub+Homepage&utm_medium=Project+tile&utm_campaign=Outlook+GIHub+Tile (accessed on 4 August 2019).
2. Mansell, P. Quantitative Survey Analysis: What Engineers and CEOs Currently Think about Sustainability and the SDGs. 2018. Available online: https://www.researchgate.net/publication/327602328_Engineers_perception_of_value_of_SDGs_and_the_current_ability_to_measure_projects'_SDG_impact (accessed on 6 June 2020).
3. Mansell, P.; Philbin, S.P.; Konstantinou, E. Using 'Creating Shared Value' to Support Measurement of the Sustainable Development Goal (SDG) Targets for Infrastructure Projects. Available online: http://www.euram-online.org/annual-conference-2019.html (accessed on 6 June 2020).
4. Elkington, J. Towards the sustainable corporation: Win-win-win business strategies for sustainable development. *Calif. Manag. Rev.* **1994**, *36*, 90–100. [CrossRef]
5. Elkington, J. Enter the triple bottom line. In *The Triple Bottom Line*; Routledge: Oxford, UK, 2013; pp. 23–38.
6. Elkington, J. 25 years ago I coined the phrase Triple bottom line. Here's why it's time to rethink it. *Harv. Bus. Rev.* **2018**, *25*, 25. Available online: https://hbr.org/2018/06/25-years-ago-i-coined-the-phrase-triple-bottom-line-heres-why-im-giving-up-on-it?utm_source=Master%20List&utm_campaign=ef4332d048-EMAIL_CAMPAIGN_2018_01_30_COPY_01&utm_medium=email&utm_term=0_458e074771-ef4332d048-449020321 (accessed on 10 July 2020).
7. Bhattacharya, A.; Oppenheim, J.; Stern, N. Driving Sustainable Development through Better Infrastructure: Key Elements of a Transformation Program. Brookings Global Working Paper Series. Available online: https://g24.org/wp-content/uploads/2016/02/Driving-Sustainable-Development-Through-Better-Infrastructure-Key-Elements-of-a-Transformation-Program-Bhattacharya-Oppenheim-Stern-July-2015.pdf (accessed on 10 July 2020).
8. The Global Commission on the Economy and Climate. Washington DC: World Resources Institute. 2014. Available online: https://www.deutsches-klima-konsortium.de/fileadmin/user_upload/pdfs/Briefings/Morgan_12_Nov_15.pdf (accessed on 10 July 2020).

9. Hartshorn, J.; Maher, M.; Crooks, J.; Stahl, R.; Bond, Z. Creative destruction: Building toward sustainability. *Can. J. Civ. Eng.* **2005**, *32*, 170–180. [CrossRef]
10. Sverdrup, H.; Rosen, K. Long-term base cation mass balances for Swedish forests and the concept of sustainability. *For. Ecol. Manag.* **1998**, *110*, 221–236. [CrossRef]
11. Costanza, R.; Patten, B.C. Defining and predicting sustainability. *Ecol. Econ.* **1995**, *15*, 193–196. [CrossRef]
12. Emas, R. The Concept of Sustainable Development: Definition and Defining Principles. Brief for GSDR 2015. Sustainable Development Goals Knowledge Platform. United Nations. 2015. Available online: https://sustainabledevelopment.un.org/content/documents/5839GSDR%202015_SD_concept_definiton_rev.pdf (accessed on 10 July 2020).
13. Moore, J.E.; Mascarenhas, A.; Bain, J.; Straus, S.E. Developing a comprehensive definition of sustainability. *Implement. Sci.* **2017**, *12*, 110. [CrossRef] [PubMed]
14. Glavič, P.; Lukman, R. Review of sustainability terms and their definitions. *J. Clean. Prod.* **2007**, *15*, 1875–1885. [CrossRef]
15. Sachs, J.D.; Schmidt-Traub, G.; Durand-Delacre, D. *Preliminary Sustainable Development Goal (SDG) Index and Dashboard*; Sustainable Development Solutions Network: New York, NY, USA, 2016.
16. Sachs, J.D.; Woo, W.T.; Yoshino, N.; Taghizadeh-Hesary, F. Importance of Green Finance for Achieving Sustainable Development Goals and Energy Security. In *Handbook of Green Finance*; Springer Science and Business Media LLC: Berlin, Germany, 2019; pp. 3–12.
17. Brundtland, G.H. *Our Common Future: Report of the World Commission on Environment and Development*; Oxford University Press: Oxford, UK, 1987.
18. Rockström, J. Wedding Cake (Donut) View of the Economy and Social Embedded as Parts of the Biosphere (Stockholm Resilience Centre). Available online: https://www.stockholmresilience.org/research/research-news/2016-06-14-how-food-connects-all-the-sdgs.html (accessed on 10 July 2020).
19. Ainger, C.M.; Fenner, R.A. *Sustainable Infrastructure: Principles into Practice*; ICE Publishing: London, UK, 2014.
20. Inter-American Development Bank, IDB. What is Sustainable Infrastructure? A Framework to Guide Sustainability across the Project Cycle. Available online: https://publications.iadb.org/publications/english/document/What_is_Sustainable_Infrastructure__A_Framework_to_Guide_Sustainability_Across_the_Project_Cycle.pdf (accessed on 10 July 2020).
21. Mansell, P.; Philbin, S.P.; Broyd, T.; Nicholson, I. Assessing the impact of infrastructure projects on global sustainable development goals. In *Proceedings of the Institution of Civil Engineers-Engineering Sustainability*; Thomas Telford Ltd.: London, UK, 2020; Volume 173, pp. 196–212. [CrossRef]
22. Global Reporting Initiative. Available online: https://www.globalreporting.org/Pages/default.aspx (accessed on 10 July 2020).
23. Global Compact, United Nations. Reporting on SDGs. Making Global Goals Local Business. Available online: https://twitter.com/globalcompact (accessed on 20 August 2019).
24. Barrett, J.R.; Peters, G.P.; Wiedmann, T.O.; Scott, K.; Lenzen, M.; Roelich, K.; Le Quéré, C. Consumption-based GHG emission accounting: A UK case study. *Clim. Policy* **2013**, *13*, 451–470. [CrossRef]
25. De Villiers, C.; Rinaldi, L.; Unerman, J. Integrated reporting: Insights, gaps and an agenda for future research. *Account. Audit. Account. J.* **2014**, *27*, 1042–1067. [CrossRef]
26. Corporate Reporting Dialogue (CRD). SDGs and the Future of Corporate Reporting. 2019. Available online: http://integratedreporting.org/wp-content/uploads/2019/02/The-Sustainable-Development-Goals-and-the-future-of-corporate-reporting-1.pdf (accessed on 19 March 2019).
27. Scott, L.; McGill, A. From Promise to Reality: Does Business Really Care about the SDGs? London: PwC. 2018. Available online: https://www.pwc.com/gx/en/sustainability/SDG/sdg-reporting-2018.pdf (accessed on 10 July 2020).
28. Global Reporting Initiative. Carrots and Sticks, Global Trends in Sustainability Reporting Regulation and Policy'. 2016. Available online: https://www.carrotsandsticks.net/wp-content/uploads/2016/05/Carrots-Sticks-2016.pdf (accessed on 10 July 2020).
29. Building Research Establishment (BRE). Available online: https://www.breeam.com/engage/research-and-development/consultation-engagement/ceequal-v6-consultation/ (accessed on 26 August 2019).
30. Pearce, O.J.; Murry, N.J.; Broyd, T.W. Halstar: Systems engineering for sustainable development. In *Proceedings of the Institution of Civil Engineers-Engineering Sustainability*; Thomas Telford Ltd.: London, UK, 2012; Volume 165, pp. 129–140.

31. McGregor, I.M.; Roberts, C. Using the SPeAR TM assessment tool in sustainable master planning. In Proceedings of the US Green Building Conference, Pittsburgh, PA, USA, 12–14 November 2003.
32. Siew, R.Y.J.; Balatbat, M.C.A.; Carmichael, D.G. A Review of building/infrastructure sustainability reporting tools (SRTs). *Smart Sustain. Built Environ.* **2013**, *2*, 106–139. [CrossRef]
33. International Organisation for Standardization (ISO). ISO 14000 Family—Environmental Management. 2019. Available online: https://www.iso.org/iso-14001-environmental-management.html (accessed on 21 March 2019).
34. Gasparatos, A. Embedded value systems in sustainability assessment tools and their implications. *J. Environ. Manag.* **2010**, *91*, 1613–1622. [CrossRef] [PubMed]
35. Awadh, O. Sustainability and green building rating systems: LEED, BREEAM, GSAS and estidama critical analysis. *J. Build. Eng.* **2017**, *11*, 25–29. [CrossRef]
36. Shivakumar, S.; Pedersen, T.; Wilkins, S.; Schuster, S. Envision: A measure of infrastructure sustainability. In *Pipelines 2014*; ASCE: Reston, VA, USA, 2014; pp. 2249–2256.
37. Infrastructure Sustainability Council of Australia (ISCA). Available online: https://www.isca.org.au/is_ratings (accessed on 24 March 2019).
38. Clevenger, C.M.; Ozbek, M.E.; Simpson, S. Review of sustainability rating systems used for infrastructure projects. In Proceedings of the 49th ASC Annual International Conference Proceedings, San Luis Obispo, CA, USA, 10–13 April 2013; pp. 10–13.
39. Butler, D.; Farmani, R.; Fu, G.; Ward, S.; Diao, K.; Imani, M. A new approach to urban water management: Safe and sure. *Procedia Eng.* **2014**, *89*, 347–354. [CrossRef]
40. Sakamoto, K. Toward a Sustainability Appraisal Framework for Transport. Available online: https://think-asia.org/bitstream/handle/11540/1417/sdwp-031.pdf?sequence=1 (accessed on 24 March 2019).
41. United Nations. *Transforming Our World: The 2030 Agenda for Sustainable Development. Resolution Adopted by the General Assembly*; United Nations: New York, NY, USA, 2015.
42. United Nations. The Sustainable Development Goals Report. Available online: https://unstats.un.org/sdgs/files/report/2018/TheSustainableDevelopmentGoalsReport2018-EN.pdf (accessed on 15 July 2019).
43. United Nations. *Intergovernmental Panel on Climate Change (IPCC). Global Warming of 1.5 °C, an IPCC Special Report on the Impacts of Global Warming of 1.5 °C above Pre-Industrial Levels and Related Global Greenhouse Gas Emission Pathways, in the Context of Strengthening the Global Response to the Threat of Climate Change, Sustainable Development, and Efforts to Eradicate Poverty*; United Nations: New York, NY, USA, 2018.
44. Inter-agency and Expert group on SDG Indicators (IAEG-SDGs). Resolution Adopted by the General Assembly on Work of the Statistical Commission Pertaining to the 2030 Agenda for Sustainable Development (A/RES/71/313). Available online: https://undocs.org/A/RES/71/313 (accessed on 6 June 2020).
45. Mansell, P.; Philbin, S.P.; Plodowski, A. Why Project Management is Critical to Achieving the SDGs, and How This Can be Achieved. Available online: https://pmcongress2019.org/ (accessed on 6 June 2020).
46. United Nations Office for Project Services (UNOPS). Infrastructure: Underpinning Sustainable Development. UNOPS, Copenhagen, Denmark. Available online: https://www.itrc.org.uk/wp-content/PDFs/ITRC-UNOPS-Infrastructure_Underpining_Sustainable%20Development.pdf (accessed on 6 June 2020).
47. Office of National Statistics (ONS). Available online: https://sustainabledevelopment-uk.github.io/reporting-status/ (accessed on 6 June 2020).
48. Morris, P.W.G. *Reconstructing Programme Management*; John Wiley & Sons: Chichester, UK, 2013.
49. Cooke-Davies, T. The "real" success factors on programmes. *Int. J. Progr. Manag.* **2007**, *20*, 185–190. [CrossRef]
50. HM Treasury. The Green Book—Central Government Guidance on Appraisal and Evaluation. 2013. Available online: https://assets.publishing.service.gov.uk/government/uploads/system/uploads/attachment_data/file/685903/The_Green_Book.pdf (accessed on 6 June 2020).
51. Dudwick, N.; Kuehnast, K.; Jones, V.N.; Woolcock, M. *Analyzing Social Capital in Context. A Guide to Using Qualitative Methods and Data*; World Bank Institute: Washington, DC, USA, 2006; pp. 1–46.
52. Thiry, M. Value management. In *Guide to Managing Projects*; Morris, P., Pinto, J., Eds.; Wiley: Holbroken, NJ, USA, 2004.
53. Müller, R.; Jugdev, K. Critical success factors in projects. *Int. J. Manag. Proj. Bus.* **2012**, *5*, 757–775. [CrossRef]

54. Weiss, C.H. Nothing as practical as good theory: Exploring theory-based evaluation for comprehensive community initiatives for children and families. In *New Approaches to Evaluating Community Initiatives: Concepts, Methods, and Contexts*; Aspen Institute: Washington, DC, USA, 1995; pp. 65–92.
55. Stein, D.; Valters, C. Understanding Theory of Change in International Development. 2012. Available online: http://www.theoryofchange.org/wp-content/uploads/toco_library/pdf/UNDERSTANDINGTHEORYOFChangeSteinValtersPN.pdf (accessed on 6 June 2020).
56. Porter, M.E. Advantage creating and sustaining superior performance. In *Competitive Advantage*; Harvard Busines School Publishing: Brighton, MA, USA, 1985.
57. Porter, M.E.; Kramer, M.R. The Big Idea: Creating Shared Value, Rethinking Capitalism. *Harv. Bus. Rev.* **2011**, *89*, 62–67.
58. Hall, J.W.; Tran, M.; Hickford, A.J.; Nicholls, R.J. *The Future of National Infrastructure: A System of Systems Approach*; Cambridge University Press: Cambridge, UK, 2016.
59. Hall, R.P.; Ranganathan, S.; Gc, R.K. A general micro-level modeling approach to analyzing interconnected SDGs: Achieving SDG 6 and more through multiple-use water services (MUS). *Sustainability* **2017**, *9*, 314. [CrossRef]
60. Thacker, S.; Hall, J. *Engineering for Sustainable Development*; Infrastructure Transition Research Consortium (ITRC), University of Oxford: Oxford, UK, 2018.
61. Thacker, S.; Adshead, D.; Fay, M.; Hallegatte, S.; Harvey, M.; Meller, H.; O'Regan, N.; Rozenberg, J.; Watkins, G.; Hall, J.W. Infrastructure for sustainable development. *Nat. Sustain.* **2019**, *2*, 324–331. [CrossRef]
62. Adams, C.A. The Sustainable Development Goals, Integrated Thinking and the Integrated Report. 2017. Available online: http://tsss.ca/inside-the-csr-report (accessed on 6 June 2020).
63. May, T. *Qualitative Research in Action*; University of Sheffield: Sheffield, UK, 2002.
64. Bhaskar, R. *A Realist Theory of Science*; Routledge: Oxford, UK, 2013.
65. Linsley, P.; Howard, D.; Owen, S. The construction of context-mechanisms-outcomes in realistic evaluation. *Nurse Res.* **2015**, *22*, 28–34. [CrossRef]
66. Pawson, R.; Tilley, N.; Tilley, N. *Realistic Evaluation*; Sage Publications: Thousand Oaks, CA, USA, 1997.
67. Pawson, R.; Tilley, N. Realistic evaluation bloodlines. *Am. J. Eval.* **2001**, *22*, 317–324. [CrossRef]
68. Sayer, A. Why critical realism? *Crit. Realis. Appl. Organ. Manag. Stud.* **2004**, *11*, 6–20.
69. Astbury, B.; Leeuw, F.L. Unpacking black boxes: Mechanisms and theory building in evaluation. *Am. J. Evaluat.* **2010**, *31*, 363–381. [CrossRef]
70. Tilley, N. EMMIE and engineering: What works as evidence to improve decisions? *Evaluation* **2016**, *22*, 304–322. [CrossRef]
71. Green Construction Board. *Three Years on Report—Reducing Carbon Reduces Cost*; Leeds City Council: Leeds, UK, 2015.
72. Anglian Water. Responsible Business. Annual Integrated Report 2018. Available online: https://www.anglianwater.co.uk/siteassets/household/about-us/annual-intergrated-report-2018.pdf (accessed on 6 June 2020).
73. Griggs, D.; Stafford-Smith, M.; Gaffney, O.; Rockström, J.; Öhman, M.C.; Shyamsundar, P.; Steffen, W.; Glaser, G.; Kanie, N.; Noble, I. Policy: Sustainable development goals for people and planet. *Nature* **2013**, *495*, 305–307. [CrossRef]
74. Organisation for Economic Co-Operation and Development (OECD). G20/OECD Principles of Corporate Governance. Paris. Available online: https://www.oecd-ilibrary.org/governance/g20-oecd-principles-of-corporate-governance-2015_9789264236882-en (accessed on 6 June 2020).
75. Anglian Water. New Models for Collaborative Working. A Guide to Community Regeneration in Wisbech—An Anglian Water Perspective. Available online: https://www.bitc.org.uk/sites/default/files/bitc_lr_wisbech_v1.compressed.pdf (accessed on 6 June 2020).
76. Kotter, J.P. *Leading Change*; Harvard Business Press: Brighton, MA, USA, 2012.

sustainability MDPI

Review

Information Technology for Business Sustainability: A Literature Review with Automated Content Analysis

Doroteja Vidmar *, Marjeta Marolt and Andreja Pucihar

Faculty of Organizational Sciences, University of Maribor, 4000 Kranj, Slovenia; marjeta.marolt@um.si (M.M.); andreja.pucihar@um.si (A.P.)
* Correspondence: doroteja.vidmar@um.si; Tel.: +386-4-2374281

Abstract: An extremely dynamic and fast-moving environment is pushing enterprises to continuous innovation and change. Managing sustainability in a digitalized environment seems to be of central importance for policy makers, as information technologies (IT), in combination with sustainability objectives, offer a wide range of opportunities for positive change. Through a systematic literature review and the application of automated content analysis, this study aims to provide insights into the latest research in the interdisciplinary field of sustainable business models and information systems. The results of the analysis, combined with a researcher's perspective, suggest that IT, which can be used to achieve sustainability objectives, are already in place and have an infinite number of potential implications in the future. The results suggest that positive economic, social, and environmental changes can be achieved by using IT as long as they are used to identify unsustainable actions and enable positive change. The analysis of research trends revealed a discrepancy between the research in the European Union and the rest of the world and pointed to several avenues for future research.

Keywords: information technology; enterprise; business model; sustainability; business sustainability; sustainable business model; IT; IS; BM; SBM

Citation: Vidmar, D.; Marolt, M.; Pucihar, A. Information Technology for Business Sustainability: A Literature Review with Automated Content Analysis. *Sustainability* **2021**, *13*, 1192. https://doi.org/10.3390/su13031192

Academic Editor: Prescott C. Ensign
Received: 30 November 2020
Accepted: 20 January 2021
Published: 23 January 2021

1. Introduction

Over the last two decades, business models (BM) have become an important research topic. However, it is only in recent years that research has underlined the importance of implementing sustainable development goals through the development and innovation of BM [1]. These so-called sustainable business models (SBM) address sustainability issues by creatively integrating eco-efficient and eco-effective innovations into existing value creation, value delivery, and value capture elements of a BM [2]. Stubbs and Cocklin [3] conceptualized SBM by bringing together fields of organizational sustainability and BM. SBM then quickly gained momentum as a field of research [4,5] and attracted researchers from various disciplines [4,6–10], e.g., environmental sciences, social sciences, engineering, computer science, mathematics, and medicine [1].

To uncover new ways for value creation, value delivery, and value capture elements of BM, several authors see the potential in emerging information technologies (IT) [11]. Chesbrough [12] was the first to point out the link between IT and BM. Subsequently, a number of research papers focused on the role of IT in reshaping BM [2,13–20]. Researchers in the area of information systems (IS) have discussed not only the contributions of IS to business value [21] but also its impact on sustainability [22]. A turning point seems to have occurred in 2010, when several authors argued for the involvement of IS in pursuit of business sustainability [23,24]. The first years of sustainability research in IS focused exclusively on reduced resource consumption (e.g., saving energy, paper, and ink), a small segment of environmental sustainability, now known as the Green IS field [25–27]. This marks an important development in business sustainability that has influenced the way enterprises around the world operate. From early observations, some researchers argued that Green IS should not only focus on the environmental impacts of corporate performance but at least also on the indirect social and economic impacts [22,28–30].

Nowadays, business environment is extremely dynamic. Digital maturity and the use of digital innovation are crucial for enterprises to successfully navigate pressures from customers, competitors, and policy makers [31–33]. The use of IT to innovate business practices through information, automation, and transformation is well documented [22,34]. Since IT can be used to enable capabilities and improve performance, the combination of IT's capabilities with sustainability objectives represent a potential to create positive changes in terms of economic, environmental, and social benefits [22,23].

Although there has been a growing interest on IT and its role in the emergence and viability of SBM among academics and practitioners in recent years [17,34–36], Nosratabadi et al. [1] argue that the focus is mainly on "sharing economy" cases and that many research topics and methodological approaches remain mainly untouched. Their comprehensive literature review [1] included work published between 2002 and 2017, and since then, several contributions have been made (e.g., [17,35,37–41]) to increase understanding of the impact of IT on SMB. Furthermore, as governing sustainability in a digitalized environment seems to be of central importance for policy makers [42,43], a comprehensive understanding of current knowledge on this topic is required. The present study, therefore, aims to provide insights into the latest research in the interdisciplinary field of SBM and IS and provide further research directions.

In accordance with this study objective, we conducted a systematic literature review using the Preferred Reporting Items for Systematic Reviews and Meta-Analyses (PRISMA) approach [44] and reviewed papers on the role of IT on SBM. The identified papers were analyzed not only manually, but also with the content analysis tool Leximancer which helped to identify and visualize key research themes to provide an understanding of the role of IT in SBM.

Our study has two contributions. First, we provided a comprehensive review of an emerging and rapidly developing interdisciplinary field that integrates the current knowledge about the role of IT in SBM. Second, we identified avenues for future investigation of this increasingly important research area.

The rest of the paper is structured as follows. In the following section, we present the methodology of the literature review. We then present the results and provided contributions to the discussion and further research directions. Finally, we provide concluding remarks and limitations of the study.

2. Methods and Data Collection

2.1. Selection of Papers

To provide an overview of current research, we first conducted a systematic literature search, using the following research terms and combining them with Boolean operators (AND and OR): business model*, sustainab*, information systems, information technolog*, and digit*. Among the publications of interest are scientific journal papers and book chapters from various disciplines. We searched the online database Web of Science.

After we obtained the first search results, we identified a total of 106 papers. Based on the recommendations of Levy and Ellis [45], we performed an additional search as follows: we searched for relevant papers by authors from the list of obtained relevant papers (eight additional papers); we searched for relevant papers by references of the obtained papers (25 additional papers), resulting in a total of 33 additional relevant papers. Based on the PRISMA statement [44], we also included four papers recommended by other sources (e.g., personalized recommendations by Mendeley, ResearchGate, and other publishers).

Following the PRISMA statement [44] and the guidelines of Kitchenham and Charters [46], we carried out an initial screening and quality assessment of the papers obtained through an initial search. Based on the review of the title and abstract, we eliminated 81 papers that were not relevant. A more detailed reading (quality assessment) of the remaining papers followed, which led to the elimination of one more paper.

At the end, we obtained a list of 61 papers, which we further examined to determine the main findings and to identify further research directions. The number of papers that we obtained through the search, assessed, and included in our content analysis is shown in the PRISMA flow diagram in Figure 1.

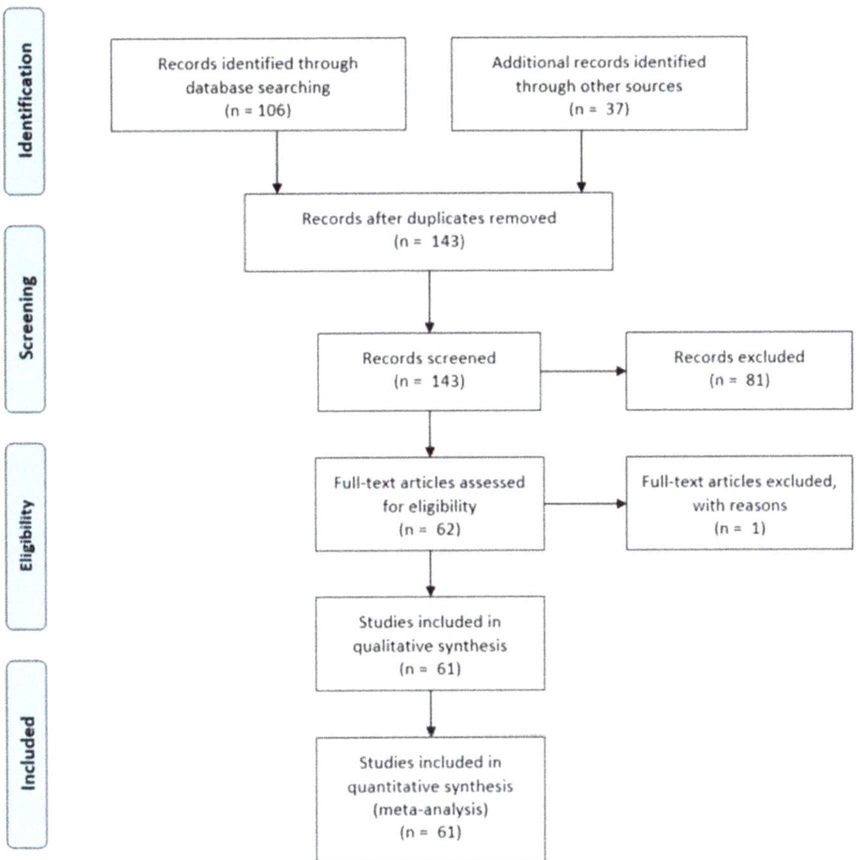

Figure 1. Selection of papers in the PRISMA flow diagram.

2.2. Analysis

While adapting the analysis process to material and research questions is challenging, immersing yourself in the data can bring many interesting points that would otherwise remain uninvestigated [47,48]. Since the available knowledge on our topic of interest is limited [49], we used qualitative content analysis to describe the phenomenon under investigation. This approach enables researchers to analyze textual material, regardless of its origin [48,50], with the aim of creating categories [50].

First, we used Tableau software [51] to analyze the number of papers per year, the journals in which they were published, and the authors' place of affiliation, to see if there are countries where research is more concentrated. Then, we continued with content analysis, combining two perspectives—that of a software program capable of quantifying and analyzing large amounts of data, and that of a human researcher capable of taking a broader perspective while looking at what is missing in the picture.

To analyze large amounts of text, we first used an automated approach to content analysis, which was performed with Leximancer [52]. The Leximancer software for automated content analysis (text analysis) that we used for our research applies Bayesian learning algorithm to break down large amounts of text into a conceivable number of relationships and categories [52,53]. From concepts and relationships, Leximancer creates "concept maps" that visualize relationships between concepts and aggregate concepts with related meanings into themes [53,54].

To ensure better results of the automated text analysis, all pdf files were first converted into text files. In addition, all unnecessary texts not related to the content were deleted, e.g., authors and their affiliation, journal name, chapter titles, tables, and captions. These files were then imported into the content analysis software Leximancer. Through several iterations of text analysis with Leximancer, we adjusted the settings of the word processor; we added standard English "stopwords" (list of common words excluded from analysis), to which we added some words, e.g., Table A1 and Figure 4. We also used the function "merge word variants", which combines concepts that have the same stems into one concept. In our case, this means that singular and plural words (e.g., model and models) are treated as one concept. We did not find any complex stemming in the results of this analysis. The results provided by Leximancer helped us to identify the main concepts, which we further elaborated through the researchers standpoint.

The results of the content analysis are presented in the following sections.

3. Results

3.1. Field Evolvement by Numbers

The literature search led to a classification of 61 papers. An analysis on an annual basis showed that the maximum number of papers was published in 2018 (Figure 2).

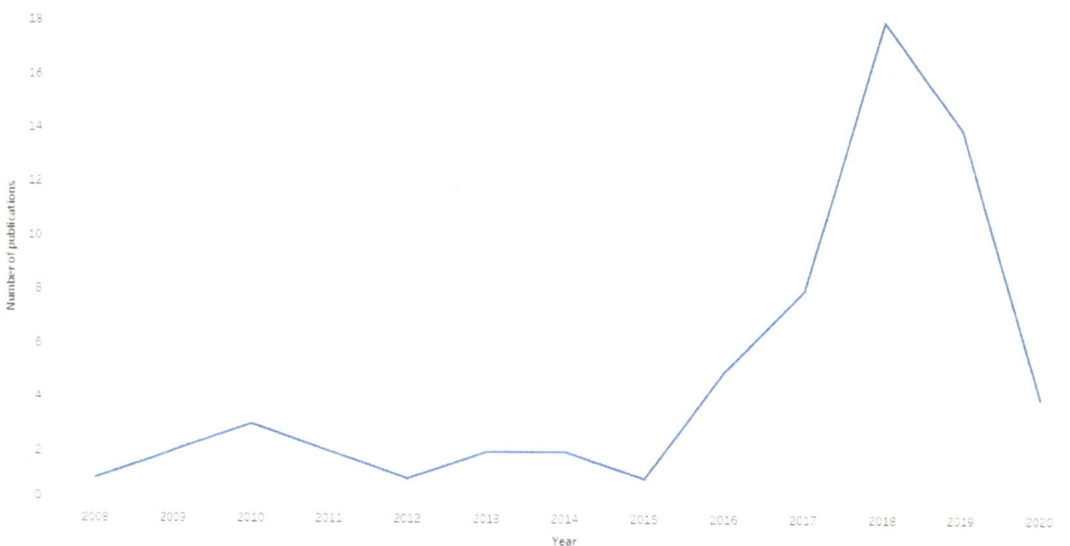

Figure 2. Number of papers per year.

The papers were also analyzed to find out in which journals they had been published over the years. Figure 3 shows that papers were published in 28 journals, the majority in the *Journal of Cleaner Production* and journal *Sustainability*.

Figure 3. Number of publications per journal and year.

The classification in Figure 4 indicates that 31 publications were theoretical, 23 qualitative, while only three were mixed methods and two quantitative. The remaining two publications are book chapters where no specific approach was applied. All included papers, methodologies as they were stated by the authors, and assigned methodological categories are listed in Table A1 in the Appendix A.

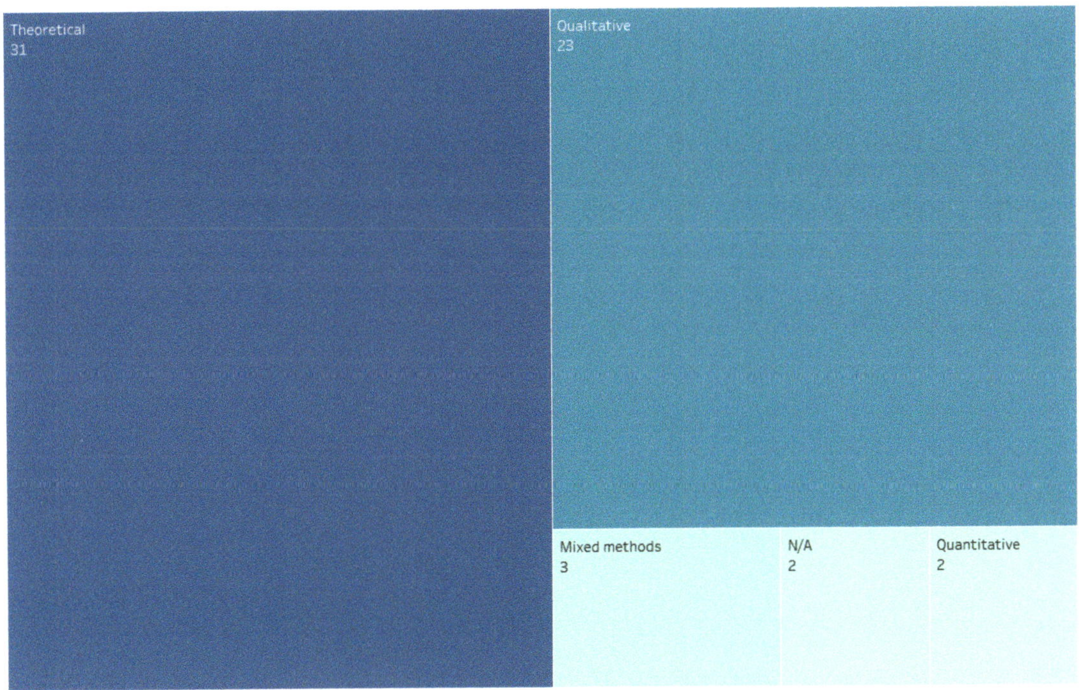

Figure 4. Article distribution by approach.

Further analysis revealed that the majority of research was conducted in European countries (Figure 5). Nevertheless, we found there are not only collaborations between authors within one country but also between different European countries, or the collaboration involves at least one researcher from outside of Europe (e.g., United States of America (USA), Canada, and Australia) (Figure 5).

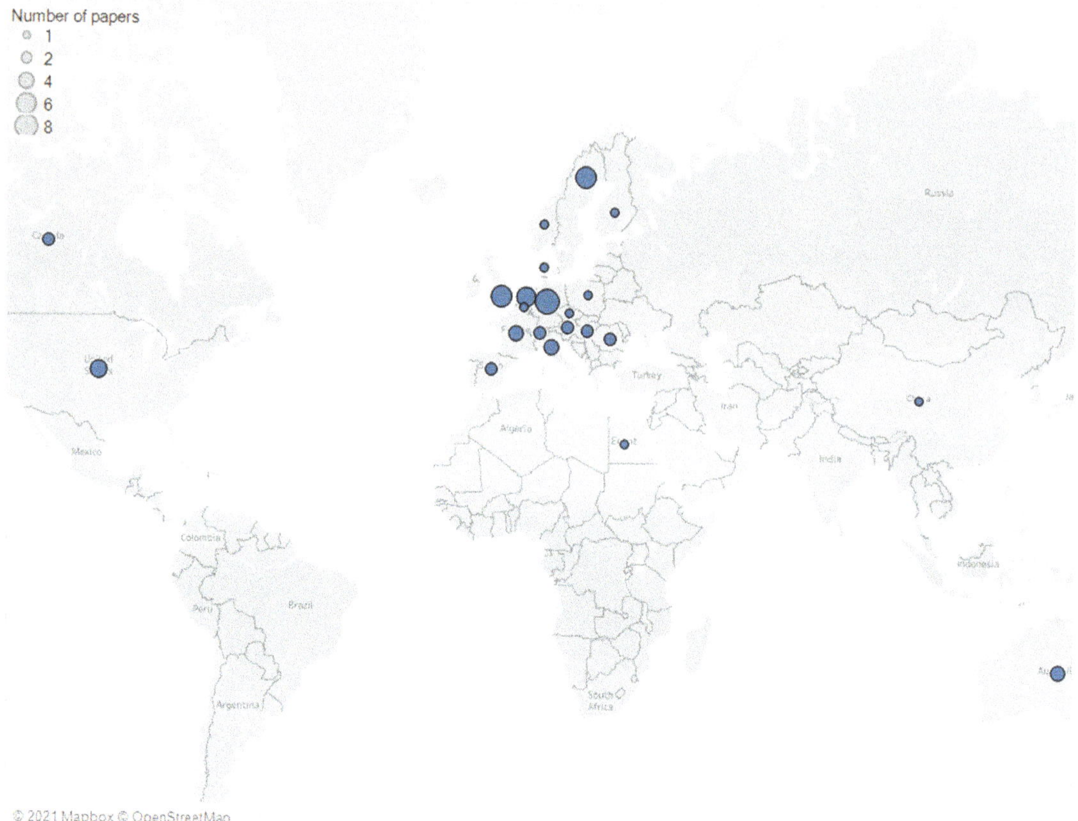

Figure 5. Number of corresponding authors per country.

3.2. Results of Content and Thematic Analysis

By analyzing 61 papers with Leximancer, we identified 14 themes (Figure 6). Themes identified by the analysis (as shown in Figure 6; the order is descending by the number of matches from the analyzed text) are "business", "sustainability", "value", "process", "research", "support", "use", "products", "customers", "effects", "information", "study", "future", and "people".

With the help of Leximancer, we also obtained a "concept map", which is shown in Figure 7. The concept map consists of themes (colored circles) and concepts that form each theme (text within the themes in black letters). The importance of themes is shown by color as a "heat map" (the brighter the theme, the more often it was found in the analyzed text) and size (the larger the theme, the more concepts were combined into it) [53,54]. The concept map also shows which themes are overlapping, e.g., in our case "business" and "sustainability"; which concepts are shared between two themes, e.g., in our case concept "innovation" lies in the overlap of themes "business" and "sustainability"; and which relationships between the concepts maintain relationships between the themes, e.g., (“business”) “model”–“innovation”–“sustainability”.

Figure 6. Identified themes.

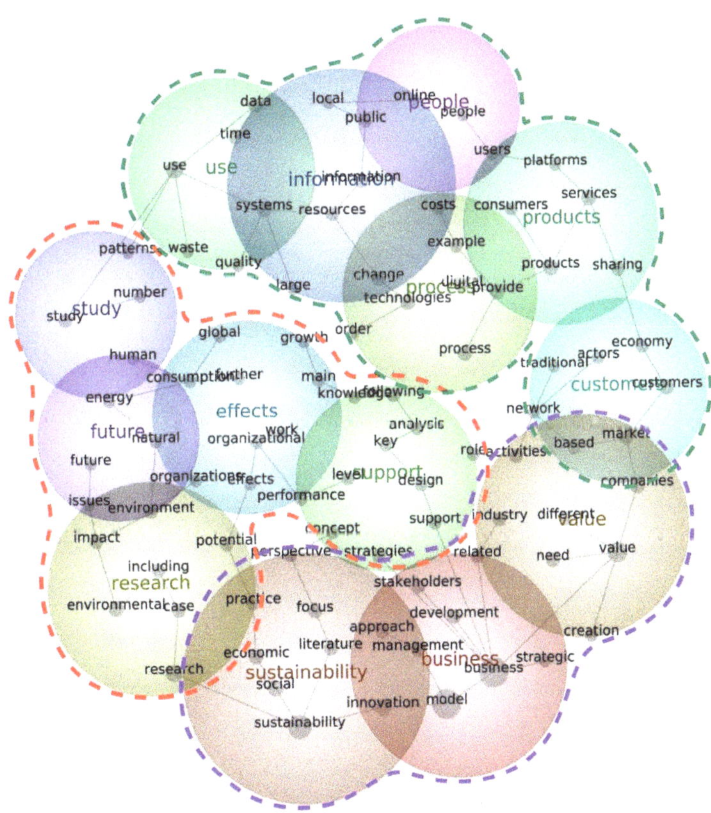

Figure 7. Leximancer concept map.

Based on our knowledge, understanding of the field, and the observations from the concept map (Figure 7), we identified the following 3 theme clusters:

The 1st cluster, which we named "Sustainable business", consists of themes "sustainability", "business", and "value". This cluster is marked with a violet dotted line. This cluster is related to value creation through SBM.

The 2nd cluster, which we named "Value creation and delivery", consists of "information", "use", "people", "process", "products", and "customers" themes. This cluster is marked with a green dotted line. The cluster is related to value creation and delivery with the use of IT in the business processes and in service and product design for the customers.

The 3rd cluster, which we named "Research", consists of "research", "future", "study", "effects", and "support" themes. This cluster is marked with a red dotted line and presents current research results and future research directions in the field of business sustainability IT.

Each cluster is thematically related to the other two clusters. The 1st cluster "Sustainable business" is connected to the 2nd cluster "value creation and delivery" through the intersection of themes "value" and "customer". More precisely, the concept "value" is connected through the concept "companies" to the concept "market". This connection can be understood as follows: "value", created by the enterprises ("companies"), is provided to the "customers" in the "market". The second connection between these two clusters is through the "network" and "market"-"based" "activities". These connections illustrate that "activities" of companies are "networked" with business partners in the "market" (ecosystem). From the broader perspective, we can also see that IT and other technologies play an important role in SBM. By using IT enterprises, we are able to collect and process a large amount of customer and other data. For instance, customer data can help identify customer needs and wants, and these insights can be used to provide value added to products and services for them. Furthermore, the production data and data from business processes can help to monitor day-to-day business tasks and processes and thus help to identify critical processes that present a threat to the environment. With this kind of knowledge, enterprises can more easily approach developing SBM.

The 2nd cluster "value creation and delivery" is connected to the 3rd cluster "research" through the themes "use", "patterns", and "study". From Figure 7, we may notice that there are not many interrelations between these two clusters. However, it is important to emphasize that the 2nd cluster also includes the use of IS in the process of value creation and delivery. Therefore, these results are not so surprising and as such support previous findings, which indicate that the multidisciplinary research of impact of IS and IT on BM and sustainability is still in its infancy.

Figure 7 shows that there are more connections between the 3rd cluster "research" and 1st cluster "sustainable business". There are five paths of connections, each indicating research on BM and sustainability. The 1st path is "sustainability", "research", "case", and "environment". This is aligned with previous research, which confirms that environmental perspective was the first focus in multidisciplinary investigation of sustainability, BM, and IS. The 2nd path is through "social", "economic", "practice", and "potential", "effects". This path shows that from a triple bottom line perspective of sustainability, research and actions should go beyond the economic perspective and should also include (previously already mentioned) environmental and social perspectives. The 3rd path illustrates a connection between "sustainability", "literature", "focus", "perspective", and "concept". The 4th path shows a connection between "business model" and "strategies", and the 5th a connection between "business model" and "support". These three paths (the 3rd, 4th, and 5th) indicate that previous research was focused on the literature review and investigation of cases (which can also be seen in Figure 4 and Table A1). Furthermore, the focus was more on the multidisciplinary field of BM and sustainability (as an important strategy and urgent need of future development of enterprises), with observable lack of IS (and IT) perspective.

In the next paragraphs, we present each cluster in detail.

3.2.1. 1st Cluster: Sustainable Business

The "sustainable business" cluster is the underlying theme of analyzed materials and represents the relationships between sustainability, BM, and value.

The text analysis showed that the strongest relationship in the analysis is between concepts "model" and "business". Concepts "business" and "model" are closely and directly connected; the strength marked in Leximancer is 5/5, as we anticipated based on the selected literature. Due to the nature of analyzed material dealing with BM, it can be assumed that in most cases this is a single concept—"business model" (BM). We confirmed that "BM" is used as one concept in most cases by examining the literature matches report. Therefore, we use "BM" as one concept throughout this paper.

"Business" and "sustainability" are the themes with the most hits from the investigated literature, closely followed by the theme "value" (Figure 6). Furthermore, the concept map (Figure 7) shows that themes "business" and "sustainability" overlap. Three concepts are shared between both themes, namely, "innovation", "management", and "approach". If we take a closer look, two paths connect themes "business" and "sustainability".

The first path, which connects the themes "business" and "sustainability", leads through the concepts "BM", "innovation", and "sustainability" and confirms that innovation of BM can lead to sustainability [14,15,34,40,55–57]. This path is also linked to the concept "value" ("value"–"BM"–"innovation"–"sustainability"), which indicates that value is used as a focal point of a BM concept by various authors [55,58–61] and in various BM frameworks [10,59,62,63]. Furthermore, it suggests that BM innovation could be the key initiator of business sustainability [2,9,55]. This indicates that when innovating BM, not only economic value but also social and environmental benefits should be taken into consideration and shared with multiple stakeholders [3,8,61,64,65].

The second path, which links the themes "business" and "sustainability", is through the concepts "BM", "management", and "approach". The entire path is classified under the theme "business", but the two concepts "management" and "approach" are shared between the two themes "business" and "sustainability". This is consistent with the observations of other researchers. For example, Jabłonski [66] stated that there are common approaches when it comes to managing BM for sustainability, including balancing stakeholders' needs and ensuring economic, environmental, and social benefits. If stakeholders value social and environmental outcomes, the value creation process should reflect this [3,61]. However, Schaltegger et al. [15] noted that business practices that lead to sustainability do not just happen, but need to be designed and deliberately and actively managed. In this cluster, the value represents a link that connects strategic goals of moving towards SBM using value creation and delivery processes that are represented in the 2nd cluster.

3.2.2. 2nd Cluster: Value Creation and Delivery

Cluster "value creation and delivery" represents the use of IS and IT in (business) processes to create value added products and services for customers. More close observation of the 2nd cluster indicates the interplay of "people" using "information" ("systems" and "technologies") in various "processes" to generate and provide value added "products" (and "services") with a particular focus on "customers" via (also online) "environment" ("sharing", "economy, and "platforms"). In this cluster, we noticed two pairs of concepts that are strongly connected and, in most cases, represent one concept. The first two concepts are "digital" and "technologies" (in a theme "process"), which, based on the strength of connections as well as in-depth investigation of the literature, represent one concept, "digital technologies". The second pair of concepts is "information" and "systems" (in the theme "information"). Additionally, in this case, based on the in-depth literature review, we understand these two concepts as a single concept "information systems".

Although we could not identify a separate theme that would include all technological (IT and IS) aspects (see Figure 7), deeper observation of the whole cluster includes a plethora of concepts related to IT and IS, for example: "digital", "technologies", "platforms", "online", "information", "systems", and "data". This could be explained by the fact that technology nowadays plays a strategic role in business. However, it is no longer considered as a separate part, but is strongly interwoven with every area and process in enterprises.

Further observation of the cluster "value creation and delivery" shows that themes "products" and "information" overlap most with other themes ("use", "people", "process", and "customer") in this cluster. Theme "products" consists of constructs "products", "services", "consumers", "sharing", and "platforms". This indicates that these concepts are very closely related to a product (and service) or value proposition. The literature suggests that proposed value today often consist of bundles of these concepts [34,38,57,67–69]. In the context of IT, products are bundled with smart sensors and/or digital services and are provided via digital platforms. If not, these products are considered to be of limited use by customers [57,67]. The direct connections between theme "products" and themes "customers", "processes", and "people" indicate that customers and users are directly involved in the development of products and services through the processes which are supported by different IT [68,70].

Theme "information" consists of concepts "information systems", "large", "resources", "change", "public", "local", and "online". This suggests that there are various ("large" numbers of) "information systems" that use different "resources" to process "information" from online "public" (e.g., open data, Internet) and local "data" (IS within enterprise). The direct connection between theme "information" and themes "use", processes", and "people" implies that "data" are generated through the "use" of "information systems" (and technologies) by "people" ("users") to support and streamline different business "processes". The implication of the use of "information systems" and "technologies" is "digital"-ization of "processes", which leads to optimization, more precisely to shorter "time" and increased "quality" of business operations ("processes"), reduced "waste" of "resources" and overall "cost" reduction (e.g., [34,35]).

Even though themes "information" and "products" are not directly connected, the connections between them through the themes "people" and "process" imply (a) "information" as a "resource" generated and used by "people" consuming "products" and related "services" on digital "platforms" [34,35] and (b) "information" as a "resource" that helps successfully incorporate "information systems" and "technologies" into enterprise "processes" with an aim to deliver innovative "products" and "services" via online experiences ("platforms") [35,67].

3.2.3. 3rd Cluster: Research

The cluster "research" represents the state of research until early 2020, future sustainability issues and organizational effects as presented in the analyzed materials, as well as the support that research can offer to enterprises.

In addition to the economic and social dimensions of sustainability, which are included with the theme "sustainability", the environmental dimension is included with the theme "research". This shows that the state of research and literature on business sustainability has historically been significantly oriented towards the environmental dimension of sustainability [24,30,71,72].

The themes "research" and "sustainability" are directly connected by the concepts "sustainability", "research", and "case", pointing to the body of knowledge that consists mainly of qualitative research case studies located in different businesses [56,57,67,68]. Theme "research" is also linked to the theme "future" through the concepts "future", "issues", "impact", and "environmental", which refer to either (a) research regarding resolving environmentally unsustainable practices in order to prevent negative consequences in the future [56] or (b) research determining scope and severity of future issues that may arise from environmentally unsustainable practices [7].

On the other side of the cluster, there is a theme "study". Theme "study" overlaps with the theme "future" and connects to the theme "use" (included in the cluster value creation and delivery). It appears that theme "study" in this case is related to research on how the use of IT-enabled sustainable solutions help enterprises achieve sustainable objectives [34,35,38,67].

4. Discussion and Further Research Directions

The aim of our study was to provide a comprehensive review of an emerging and rapidly developing interdisciplinary field and to integrate current knowledge on the role of IT in SBM. To this end, we conducted a systematic literature review of 61 papers related to the role of IT in SBM using the content analysis tool Leximancer. Based on the analysis, we identified 14 themes interrelated through various concepts (key words). Observations of the visual results, provided by Leximancer (Figure 7) helped us to gain deeper insights into the current body of knowledge in this interdisciplinary field, provide an interpretation according to our understanding (human perspective), and suggest avenues for future research.

In the remainder of this chapter, we discuss the scope and outlets of current publications and results. Furthermore, we provide avenues for future research.

4.1. Discussion of Scope and Outlets of Current Publications

The results of our literature review on IT, BM, and sustainability show that the number of contributions has been increasing in recent years. Even though the number of papers published in 2019 was lower than that of 2018, we expect the trend towards more quality and quantity of research in this interdisciplinary field to continue. At least, and above all, this can be said for the European Union (EU), where the European Commission (EC) is pushing the agenda for research on digitalization and sustainability. EC Agenda states that competitiveness in the coming years will depend on the sustainability and the ability to exploit IT [42]. However, this research should not be limited to the EU, as digitalization and sustainability are global matters that should concern enterprises and policy makers worldwide [73].

Figure 5 (number of corresponding authors per country) shows that the majority of authors of published papers are based in the EU, where policy makers and enterprises in general have a strong interest in sustainability issues. It is worrying that this could be a result of a different corporate governance structure. Most EU enterprises are governed by a two-tier board system of corporate governance [74]. A dual structure of management and supervisory board that have different roles creates opportunities for different types of values (e.g., economic, environmental, and social), while a unitary board system consisting of a single board of directors, as is common in the USA, tends to outweigh social and environmental concerns in favor of economic gains [74,75]. This is consistent with Stubbs and Cocklin [3] who argue that absentee shareholders (shareholders who are not involved in a community in which the enterprise operates) tend to focus on economic rather than social and environmental benefits.

The findings in Figure 3 (number of publications per journal and year) are supported by Parida et al. [17], who note the debate on sustainability has moved from journals on environmental management to journals on strategic management and entrepreneurship, where sustainability, innovation, and competitiveness are now the central issues. It is also notable that many journals from the IS discipline in recent years have organized Special Issues on the emerging theme of sustainability [76–78], where the central point of investigation was related to the use of IT, digitalization, and digital transformation for innovation or development of SBM.

4.2. Discussion of Findings

The IS discipline has more than five decades of evolvement (Davis, 2006). A historical view of development of the field shows that the early phase of investigation was related to electronic data interchange between organizations (up to the 1980). This era was followed by electronic business, which was enabled by a wider use of the Internet (1990 to 2005). From 2005 on (up to 2011), research was focused on electronic interactions between all stakeholders in society. The last decade (from 2011 on) was dedicated to digital transformation, which is the result of new emerging technologies (e.g., social, mobile, analytics, artificial intelligence, cloud and high performance computing, Internet of Things, and robotics) and their impact on enterprises, organizations, individuals, and society [54]. In the business context, digital transformation refers to a process of redesign or innovation of BM as a result of the adoption and use of IT, which create digital capabilities [54,79] However, in the last decade, there has also been an emerging need for another transformation—so-called sustainability transformation [35]. IT and digital transformation bring enormous opportunities to respond to this emerging need pursued not only by evidence from the environment, and expectations from citizens and customers, but also as a formal demand from governments [80,81].

It is evident that IT have become a main component of innovation and new, changed ways of value generation, delivery, and resource distribution [40,82,83]. Our findings suggest that processes coupled with digital capabilities and IT can lead to savings not only in terms of costs but also in general resource use and distribution. In addition, innovative, digital BM show that data (e.g., time and patterns of use, generated waste) automatically provided by users via IT can help identify excessive resource use and waste, e.g., excessive fuel use and need for maintenance in car sharing [34,35]. Furthermore, by changing BM in a way that charges for access to products instead of ownership [35] (e.g., car sharing), ownership is left in the hands of enterprises. Ownership provides an incentive for enterprises to create high quality products, cause less waste through the use of products, and positively influence economic and environmental sustainability [8]. Taking ownership of products out of business transactions provides lower entry costs for users; social sustainability is improved by enabling people to pay only for the actual use of products [57], and larger customer pools for products with higher added value (e.g., higher quality and sustainably sourced) are created.

According to Yang, Evans, Vladimirova, and Rana [59], identifying uncaptured value through data on waste and resource use can lead to new value opportunities and improved sustainability. Value uncaptured can be transformed into value captured faster than new value can be created (by creating products out of wasted resources or by charging for previously free services). For example, through online business processes and the online presence of products and users (e.g., websites, digital platforms, and social media), data are collected [35]. These data enable continuous business model innovation (BMI), iterative development of solutions, and rapid validation of business viability, saving time and resources in the process [59,64]. In addition, our results suggest that environmental effects, such as the reduction in waste and resource use of enterprises, can be realized through customer needs if appropriate BM is used [34]. This implies that innovative, potential SBM are enabled by the use of IT.

Existing studies are focused either on the business perspective or the customer perspective. The business perspective attempts to capture individual experiences of enterprises and provide a deeper understanding of how enterprises use IT and tackle sustainability issues. The customer perspective attempts to investigate customer use of resources or their motivation to use IT with the aim of achieving sustainable goals. Research topics include the role of customers and motivating customers by rewarding sustainable behavior. The latter is based on innovative BMs that are designed to achieve environmental and social goals, including lower resource consumption, less negative effects for the environment, inclusive models that reduce the entry price of resources by allowing customers to pay for use rather than ownership, or sharing savings with the enterprise. The customer perspective, including how to engage customers in sustainable business activities, is related to this research and is an important issue in itself [35,67].

There is no indication that a particular type of IT could be most beneficial for SBM, which suggests that SBM are evolving in line with IT advancements [17,41]. Enterprises invested in continuous BMI strive to embed state-of-the-art IT that are compatible with their existing processes, technologies, strategies, and objectives [41]. It is evident that in the future, new technologies will continue to emerge, and digital transformation will be an ongoing process in every enterprise and society. Digital transformation should be used as an enabler for the transformation of enterprises towards designing more responsible BMs, which in addition to economic also consider environmental and social dimensions of BM.

4.3. Future Research Directions

In the last decade, the importance of the sustainability perspective has already been raised by researchers in different disciplines. For example, Seidel et al. [84] urged IS researchers to integrate sustainability as an essential part of their research. Although it is evident that continuous digitalization and digital transformation of BM provide enormous opportunities for development of SBM [17,41], the interdisciplinary field of research in this area is still at its early stage of development. Researchers are still searching for a deeper understanding of how enterprises achieve sustainability objectives with the use of IT [17,66,85].

Our analyses show that in most cases, the focus of previous research was towards the effects of IT on organizational performance and work (business processes and operations) [34,57]. Another relatively well-represented focus is towards environmental sustainability [26,30,86,87]. Furthermore, our results suggest that future research should continue to focus on sustainable consumption of resources (re-use and circular economy), especially natural (e.g., water and energy), and on environmental perspectives of sustainability. Moreover, current research and practice in SBM have to date paid little attention to customer (human–social perspective) needs and their integration with IT to generate sustainable business value [64].

Results of our study support suggestions of previous research, emphasizing the need for deeper exploration of the emerging field of SBM [4,5,8], with particular focus on the impacts of IT on achieving sustainability goals [23,34,86]. In the future, interdisciplinary research on sustainability and IS will be needed for further investigation of this dynamic and fast evolving field [17,22,66,71,84,85].

Our results suggest that existing research is mainly of a qualitative nature [1,4,34,57], namely, case studies conducted in different enterprises. As the field of research is still in its early phase and of a multidisciplinary nature (and as such of higher complexity), case studies will remain an important research method. For the purpose of generalization of research results, a multimethod approach, the pursuit of novel data sources, methods, and tools to experiment with ways to reach sustainability objectives will be needed [1,4,34]. In addition, as current research from an enterprise perspective is based mostly in the EU, more research is needed to identify if there is a correlation between sustainability efforts and management system (one-tier vs. two-tier) on an organizational and national level.

With the rapid development of digital transformation and the urgent need for sustainability transformation, this interdisciplinary field will be extremely dynamic in the future from the perspective of its evolvement and research opportunities.

5. Conclusions and Limitations

The momentum of digital transformation and the rapid pace of digitization, coupled with the need for more sustainability in business, provide substantial opportunities for creating new value propositions as well as new BM [41]. Many emerging SBM are fully digitalized and heavily driven by widespread use of IT [34,67].

It is evident that unsustainable BM, driven only by economic value, has already caused observable damage to our environment as well as in society. However, in recent years, policy makers of many countries have put sustainability at the top of their agendas for future development [88,89]. For example, the European Commission (EC) [42] has already emphasized that future competitiveness will depend on the ability to exploit the opportunities of IT to move towards sustainability and resource-efficiency. Another important achievement from the policy and law perspective is related to obligatory reporting of sustainability practices for large enterprises. Namely, from 2018, large public-listed enterprises in the EU have to provide public reports about the environmental and social effects of their business practices on their employees and society [80,81]. Therefore, we may say that the overall performance of enterprises is nowadays already measured from the sustainability perspective, which includes economic but also social and environmental perspectives at least in large enterprises. However, to ensure that sustainability will become an integral part of daily business and of our lives, many regulations and (behavioral) changes will have to be implemented in the future.

Nowadays, IT are an integral part of enterprise strategy. Its role is represented in IT or digital strategy, which must be aligned with business strategy [90]. As such, IT supports operations of practically all elements of BM. The momentum of digital transformation and the rapid pace of digitalization, coupled with the need for more sustainability, provide substantial opportunities for creating new value propositions as well as new BM [41].

It is evident that the implementation of IT only to achieve higher efficiency and competitive advantage is insufficient. Much more responsible, and less exploitative, economic and BM practices are needed for the overall benefit of human beings, societies, and our natural resources and environment [54]. In the future, IT will have to be used to design solutions and BMs aligned with sustainability goals. For example, solutions will have to be made to address different societal challenges, where IT can provide new value-added services. In the context of demographic challenges, digital (care) services for the elderly, e-inclusion of the elderly, digital health solutions for citizens, etc. will have to be further designed. In the context of consumption, better planning and monitoring of food production according to the real needs, fair distribution, less waste, and other solutions can be developed. In recent years, we have observed heavy pressures on various tourism destinations and points of interest around the world. As this type of, to date, in many cases, only economically driven, BM has already caused damage in the natural environment, it is obvious that new solutions are needed to regulate (over) tourism in the future and protect natural and cultural heritage. New solutions can be related to virtual reality, mobile apps that will alert and redirect tourist to less populated points of interest, co-creation of new itineraries by providers and tourists, etc.

Since March 2019, the world has also faced the COVID-19 pandemic. While scientists provided the vaccine in only 9 months, IT played another important role in enterprises. In some industries, those enterprises, which were able to provide their employees remote access to the IS from their homes, were able to continue their business operations. On the other hand, many enterprises, which were in an earlier stage of digitalization, had to close their businesses. The momentum of the COVID-19 pandemic crises has pointed to the importance of IT like never before. In this time, many enterprises increased investments in IT and moved to digital business faster than ever before. This movement should be, from now on, permanent and continuous.

We may conclude that the results of our study provide insights into past research in the multidisciplinary field of IS and management, with particular focus on the impact of IT on SBM. Our results revealed that this multidisciplinary field of research is relatively young, however fast evolving in the last decade. New technologies will create new opportunities for digital transformation and design of digital solutions, services, models, and societies. However, these solutions will have to be created according to sustainability goals. To achieve these goals, the collaboration of all stakeholders in society will be needed (governments, enterprises, researchers, IT providers, etc.). In addition, researchers from different disciplines will have to cooperate and take an active role in these endeavors.

Although the approach with which we combined the strengths of IT and the human mind to analyze large amounts of data has its advantages, it also has shortcomings. First, the content analysis tool did not provide a definitive answer through analysis, but only a starting point. It is up to the researcher to provide a meaningful discussion supported by the literature, move from description and patterns to interpretation, determine the underlying meanings of concepts and relationships identified, and observe the gaps in the process—something that software cannot do. In other words, it is possible the results provided by the content analysis tool influenced our final judgements. In addition, Leximancer identifies single words as concepts, which means that multiword concepts cannot be identified, but are broken down into single word concepts and even placed under different themes (e.g., BM, IS, IT, and SBM). Although there are instances where it is possible to conclude that such a multiword concept is involved (e.g., overlapping concepts business and model and information systems), we found that this is a problem for content analysis in the field of information systems, as two main concepts, "information systems" and "information technology", cannot be identified. In addition, the authors frequently used the acronyms "IS" and "IT" in the articles analyzed, which cannot be treated separately from the English words "is" and "it" in Leximancer. Thus, in our content analysis, IS and IT are reflected in the results through concepts such as "information", "systems", and "technologies" as well as other related concepts, such as "digital", "data", and "platforms". Since our aim was to provide insight into the extent and ways IT and IS are involved in SBM, there may be variations in the results provided by Leximancer. However, this also represents an opportunity for further investigation in the field.

Author Contributions: Conceptualization, D.V. and A.P.; methodology, D.V.; software, D.V. and M.M.; validation, D.V., M.M. and A.P.; formal analysis, D.V.; investigation, D.V.; resources, D.V.; data curation, D.V. and M.M.; writing—original draft preparation, D.V.; writing—review and editing, D.V., M.M. and A.P.; visualization, D.V. and M.M.; supervision, A.P.; funding acquisition, D.V. and A.P. All authors have read and agreed to the published version of the manuscript.

Funding: This research was supported by the Slovenian Research Agency: Program No. P5-0018—Decision Support Systems in Digital Business and postgraduate research funding for Young Researchers No. 1000-20-0552.

Conflicts of Interest: The authors declare no conflict of interest.

Appendix A

All the selected papers that were identified and analyzed in the review with Leximancer are listed in Table A1. Table A1 includes 61 papers published from 2008 to 2020. Papers are listed in descending order by year of publishing.

Table A1 includes 6 columns: authors; year—year when a paper was published; title—full title of paper (book chapter titles are followed by book title); journal—title of journal in which paper was published (book chapters and conference papers are marked as such); type of paper (as stated by authors)—type of paper or methodological approach as authors described it in each paper; and assigned methodological category. We assigned one of four methodological categories to each paper for clarity and in order to be able to visually present the methodological approach. We categorized all papers except book chapters (marked N/A) into four methodological categories.

The assigned methodological categories are: theoretical (literature reviews, introduction to Special Issue, editorial, overview, framework development, and scientometric analysis), qualitative (case study, experimental design, and framework development based on or tested with case studies), quantitative (survey), mixed methods (combination of qualitative and quantitative—usually case study and survey, also when preceded by a literature review).

Table A1. Selected papers that were analyzed in the review.

Authors	Year	Title	Journal	Type of Paper (as Stated by Authors)	Assigned Methodological Category
Bocken, Nancy, Lisa Smeke Morales, and Matthias Lehner	2020	"Sufficiency Business Strategies in the Food Industry—The Case of Oatly"	*Sustainability*	Literature and practice review, case study	Qualitative
ElMassah, Suzanna, and Mahmoud Mohieldin	2020	"Digital Transformation and Localizing the Sustainable Development Goals (SDGs)"	*Ecological Economics*	Case study	Qualitative
Buda, Gabriella, Barbara Pethes, and Jozsef Lehota	2020	"Dominant Consumer Attitudes in the Sharing Economy—A Representative Study in Hungary"	*Resources*	Survey	Quantitative
Gil-Gomez, Hermenegildo, Vicente Guerola-Navarro, Raul Oltra-Badenes, and José Antonio Lozano-Quilis	2020	"Customer Relationship Management: Digital Transformation and Sustainable Business Model Innovation"	*Economic Research-Ekonomska Istraživanja*	Literature review	Theoretical
Bocken, Nancy	2019	"Sustainable Consumption through New Business Models: The Role of Sustainable Entrepreneurship" In Sustainable Entrepreneurship: Discovering, Creating and Seizing Opportunities for Blended Value Generation, edited by A. Lindgreen, F. Maon, and C. Vallaster.	Book chapter	Book chapter—exploration through illustrative cases	N/A
Delgado-de Miguel, Juan-Francisco, Tamar Buil-López Menchero, Miguel-Ángel Esteban-Navarro, and Miguel-Ángel García-Madurga	2019	"Proximity Trade and Urban Sustainability: Small Retailers' Expectations Towards Local Online Marketplaces"	*Sustainability*	Semi-structured in-depth interviews	Qualitative
Ievoli, Corrado, Angelo Belliggiano, Danilo Marandola, Pierluigi Milone, and Pierluigi Ventura	2019	"Information and Communication Infrastructures and New Business Models in Rural Areas: The Case of Molise Region in Italy"	*European Countryside*	Case study	Qualitative
Ockwell, David, Joanes Atela, Kennedy Mbeva, Victoria Chengo, Rob Byrne, Rachael Durrant, Victoria Kasprowicz, and Adrian Ely	2019	"Can Pay-As-You-Go, Digitally Enabled Business Models Support Sustainability Transformations in Developing Countries? Outstanding Questions and a Theoretical Basis for Future Research"	*Sustainability*	Literature review, workshops, interviews	Qualitative
Olah, Judit, Nicodemus Kitukutha, Hossam Haddad, Miklos Pakurar, Domician Mate, and Jozsef Popp	2019	"Achieving Sustainable E-Commerce in Environmental, Social and Economic Dimensions by Taking Possible Trade-Offs"	*Sustainability*	Literature review and case study	Qualitative
Vorraber, Wolfgang, and Matthias Müller	2019	"A Networked Analysis and Engineering Framework for New Business Models"	*Sustainability*	Framework development, case study	Qualitative

Table A1. *Cont.*

Authors	Year	Title	Journal	Type of Paper (as Stated by Authors)	Assigned Methodological Category
Dumitriu, Dan, Gheorghe Militaru, Dana Corina Deselnicu, Andrei Niculescu, and Mirona Ana-Maria Popescu	2019	"A Perspective Over Modern SMEs: Managing Brand Equity, Growth and Sustainability Through Digital Marketing Tools and Techniques"	*Sustainability*	Survey	Quantitative
Denuwara, Navodya, Juha Maijala, and Marko Hakovirta	2019	"Sustainability Benefits of RFID Technology in the Apparel Industry"	*Sustainability*	Literature review	Theoretical
Freudenreich, Birte, Florian Lüdeke-Freund, and Stefan Schaltegger	2019	"A Stakeholder Theory Perspective on Business Models: Value Creation for Sustainability"	*Journal of Business Ethics*	Framework development	Theoretical
Gössling, Stefan, and Michael Hall	2019	"Sharing versus Collaborative Economy: How to Align ICT Developments and the SDGs in Tourism?"	*Journal of Sustainable Tourism*	Discourse analysis and literature review	Theoretical
Jose, Charbel, Chiappetta Jabbour, Ana Beatriz Lopes De Sousa Jabbour, Joseph Sarkis, and Godinho Filho	2019	"Unlocking the Circular Economy through New Business Models Based on Large-Scale Data: An Integrative Framework and Research Agenda"	*Technological Forecasting and Social Change*	Framework development	Theoretical
Parida, Vinit, and Joakim Wincent	2019	"Why and How to Compete through Sustainability: A Review and Outline of Trends Influencing Firm and Network-Level Transformation"	*International Entrepreneurship and Management Journal*	Literature review, introduction to Special Issue	Theoretical
Parida, Vinit, David Sjödin, and Wiebke Reim	2019	"Reviewing Literature on Digitalization, Business Model Innovation, and Sustainable Industry: Past Achievements and Future Promises"	*Sustainability*	Literature review, introduction to Special Issue, framework development	Theoretical
Moro Visconti, Roberto, and Donato Morea	2019	"Big Data for the Sustainability of Healthcare Project Financing"	*Sustainability*	Simulations based on empirical cases	Qualitative
Bouwman, Harry, Shahrokh Nikou, Francisco J Molina-Castillo, and Mark De Reuver	2018	"The Impact of Digitalization on Business Models"	*Digital Policy, Regulation and Governance*	Survey and case study	Mixed methods
Jabłonski, Marek	2018	"Value Migration to the Sustainable Business Models of Digital Economy Companies on the Capital Market"	*Sustainability*	Mixed methods (survey and analysis of available documents)	Mixed methods
Yip, Angus W. H., and Nancy Bocken	2018	"Sustainable Business Model Archetypes for the Banking Industry"	*Journal of Cleaner Production*	Literature and practice review, semi-structured interviews, survey	Mixed methods

Table A1. *Cont.*

Authors	Year	Title	Journal	Type of Paper (as Stated by Authors)	Assigned Methodological Category
Lüdeke-Freund, Florian, René Bohnsack, Henning Breuer, and Lorenzo Massa	2018	"Research on Sustainable Business Model Patterns: Status Quo, Methodological Issues, and a Research Agenda" In Sustainable Business Models, edited by A. Aagaard.	Book chapter	Book chapter	N/A
Bertola, Paola, and Jose Teunissen	2018	"Fashion 4.0. Innovating Fashion Industry through Digital Transformation"	*Research Journal of Textile and Apparel*	Positioning essay enriched by case studies	Qualitative
Bressanelli, Gianmarco, Federico Adrodegari, Marco Perona, and Nicola Saccani	2018	"Exploring How Usage-Focused Business Models Enable Circular Economy through Digital Technologies"	*Sustainability*	Framework development based on literature and case study	Qualitative
Hildebrandt, Björn, Andre Hanelt, and Sebastian Firk	2018	"Sharing Yet Caring: Mitigating Moral Hazard in Access-Based Consumption through IS-Enabled Value Co-Capturing with Consumers"	*Business and Information Systems Engineering*	Quasi-experimental research design based on a case study	Qualitative
Piscicelli, Laura, Geke D S Ludden, and Tim Cooper	2018	"What Makes a Sustainable Business Model Successful? An Empirical Comparison of Two Peer-to-Peer Goods-Sharing Platforms"	*Journal of Cleaner Production*	Case study	Qualitative
Pohludka, Michal, Hana Stverkova, and Beata Ślusarczyk	2018	"Implementation and Unification of the ERP System in a Global Company as a Strategic Decision for Sustainable Entrepreneurship"	*Sustainability*	Case study	Qualitative
Rajala, Risto, Esko Hakanen, Juri Matilla, Timo Seppälä, and Mika Westerlund	2018	"How Do Intelligent Goods Shape Closed-Loop Systems?"	*California Management Review*	Case study	Qualitative
Bieser, Jan C. T., and Lorenz M. Hilty	2018	"Assessing Indirect Environmental Effects of Information and Communication Technology (ICT): A Systematic Literature Review"	*Sustainability*	Literature review	Theoretical
Brenner, Barbara	2018	"Transformative Sustainable Business Models in the Light of the Digital Imperative—A Global Business Economics Perspective'	*Sustainability*	Framework development based on literature and practice review	Theoretical
Camacho-Otero, Juana, Casper Boks, and Ida Nilstad Pettersen	2018	'Consumption in the Circular Economy: A Literature Review"	*Sustainability*	Literature review	Theoretical

Table A1. *Cont.*

Authors	Year	Title	Journal	Type of Paper (as Stated by Authors)	Assigned Methodological Category
Dentchev, Nikolay, Romana Rauter, Lára Jóhannsdóttir, Yuliya Snihur, Michele Rosano, Rupert Baumgartner, Timo Nyberg, Xingfu Tang, Bart van Hoof, and Jan Jonker	2018	"Embracing the Variety of Sustainable Business Models: A Prolific Field of Research and a Future Research Agenda"	*Journal of Cleaner Production*	Overview—an introduction to the Special Issue	Theoretical
Jin, Yuran, and Shoufeng Ji	2018	"Mapping Hotspots and Emerging Trends of Business Model Innovation under Networking in Internet of Things"	*Journal on Wireless Communications and Networking*	Scientometric analysis	Theoretical
Shetty, Vivek, John Yamamoto, and Kenneth Yale	2018	"Re-Architecting Oral Healthcare for the 21st Century"	*Journal of Dentistry*	Overview (methodology not stated in the paper)	Theoretical
Geissdoerfer, Martin, Doroteya Vladimirova, and Steve Evans	2018	"Sustainable Business Model Innovation: A Review"	*Journal of Cleaner Production*	Literature review	Theoretical
Mountean, Mihaela	2018	"Business Intelligence Issues for Sustainability Projects".	*Sustainability*	Multi-dimensional data modeling	Theoretical
Baldassarre, B., G. Calabretta, Nancy Bocken, and T. Jaskiewicz	2017	"Bridging Sustainable Business Model Innovation and User-Driven Innovation: A Process for Sustainable Value Proposition Design"	*Journal of Cleaner Production*	Research through design	Qualitative
Hanelt, Andre, Sebastian Busse, and Lutz M. Kolbe	2017	"Driving Business Transformation toward Sustainability: Exploring the Impact of Supporting IS on the Performance Contribution of Eco-Innovations"	*Information Systems Journal*	Case study	Qualitative
Moreno, Mariale, Christopher Turner, Ashutosh Tiwari, Windo Hutabarat, Fiona Charnley, Debora Widjaja, and Luigi Mondini	2017	"Re-Distributed Manufacturing to Achieve a Circular Economy: A Case Study Utilizing IDEF0 Modeling"	*Conference proceedings*	Case study	Qualitative
Yang, M, S Evans, D Vladimirova, and P Rana	2017	"Value Uncaptured Perspective for Sustainable Business Model Innovation"	*Journal of Cleaner Production*	Framework development, case study	Qualitative
Evans, Steve, Doroteya Vladimirova, Maria Holgado, Kirsten Van Fossen, Miying Yang, Elisabete A. Silva, and Claire Y. Barlow	2017	"Business Model Innovation for Sustainability: Towards a Unified Perspective for Creation of Sustainable Business Models"	*Business Strategy and the Environment*	Literature review	Theoretical

Table A1. *Cont.*

Authors	Year	Title	Journal	Type of Paper (as Stated by Authors)	Assigned Methodological Category
Lüdeke-Freund, Florian, and Krzysztof Dembek	2017	"Sustainable Business Model Research and Practice: Emerging Field or Passing Fancy?"	*Journal of Cleaner Production*	Analysis of emerging research field	Theoretical
Pagoropoulos, Aris, Daniela C. A. Pigosso, and Tim C. McAloone	2017	"The Emergent Role of Digital Technologies in the Circular Economy: A Review"	*Conference proceedings*	Literature review	Theoretical
Seele, Peter, and Irina Lock	2017	"The Game-Changing Potential of Digitalization for Sustainability: Possibilities, Perils, and Pathways"	*Sustain Sci*	Editorial, introduction to Special Issue	Theoretical
Bocken, Nancy, and S. W. Short	2016	"Towards a Sufficiency-Driven Business Model: Experiences and Opportunities"	*Environmental Innovation and Societal Transitions*	Case study	Qualitative
Breidbach, Christoph F., and Paul P. Maglio	2016	"Technology-Enabled Value Co-Creation: An Empirical Analysis of Actors, Resources, and Practices"	*Industrial Marketing Management*	Case study	Qualitative
Heiskala, Mikko, Jani-Pekka Jokinen, and Markku Tinnilä	2016	"Crowdsensing-Based Transportation Services—An Analysis from Business Model and Sustainability Viewpoints"	*Research in Transportation Business and Management*	Case study	Qualitative
Davidsson, Paul, Banafsheh Hajinasab, Johan Holmgren, Åse Jevinger, and Jan Persson	2016	"The Fourth Wave of Digitalization and Public Transport: Opportunities and Challenges"	*Sustainability*	Analysis of literature and explorative studies	Theoretical
Schaltegger, Stefan, Erik G Hansen, and Florian Lüdeke-Freund	2016	"Business Models for Sustainability: Origins, Present Research, and Future Avenues"	*Organization and Environment*	Editorial, introduction to Special Issue	Theoretical
Upward, Anthony, and Peter H. Jones. 2015	2015	"An Ontology for Strongly Sustainable Business Models: Defining an Enterprise Framework Compatible with Natural and Social Science"	*Organization and Environment*	Framework development	Theoretical
Bocken, Nancy, S. W. Short, P. Rana, and S. Evans	2014	"A Literature and Practice Review to Develop Sustainable Business Model Archetypes"	*Journal of Cleaner Production*	Literature and practice review	Theoretical
Bohnsack, René, Jonatan Pinkse, and Ans Kolk	2014	"Business Models for Sustainable Technologies: Exploring Business Model Evolution in the Case of Electric Vehicles"	*Research Policy*	Content analysis	Theoretical

Table A1. *Cont.*

Authors	Year	Title	Journal	Type of Paper (as Stated by Authors)	Assigned Methodological Category
Boons, Frank, and Florian Lüdeke-Freund	2013	"Business Models for Sustainable Innovation: State-of-the-Art and Steps towards a Research Agenda"	*Journal of Cleaner Production*	Literature review	Theoretical
Boons, Frank, Carlos Montalvo, Jaco Quist, and Marcus Wagner	2013	"Sustainable Innovation, Business Models and Economic Performance: An Overview"	*Journal of Cleaner Production*	Overview—an introduction to the Special Issue	Theoretical
Schaltegger, Stefan, Florian Lüdeke-Freund, and Erik G Hansen	2012	"Business Cases for Sustainability The Role of Business Model Innovation for Corporate Sustainability"	*International Journal of Innovation and Sustainable Development*	Framework development	Theoretical
Dao, Viet, Ian Langella, and Jerry Carbo	2011	"From Green to Sustainability: Information Technology and an Integrated Sustainability Framework"	*Journal of Strategic Information Systems*	Framework development	Theoretical
Elliot, Steve	2011	"Transdisciplinary Perspectives on Environmental Sustainability: A Resource Base and Framework for IT-Enabled Business Transformation"	*MIS Quarterly*	Literature review and framework development	Theoretical
Doz, Yves L, and Mikko Kosonen	2010	"Embedding Strategic Agility A Leadership Agenda for Accelerating Business Model Renewal"	*Long Range Planning*	Proposition of actions for renewal and transformation of business models based on practice review	Qualitative
Lüdeke-Freund, Florian	2010	"Towards a Conceptual Framework of 'Business Models for Sustainability"	*Conference proceedings*	Framework development	Theoretical
Melville, Nigel P.	2010	"Information Systems Innovation for Environmental Sustainability"	*MIS Quarterly*	Development of research agenda and framework	Theoretical
Stubbs, Wendy, and Chris Cocklin	2008	"Conceptualizing a 'Sustainability Business Model'"	*Organization and Environment*	Case study, framework development	Qualitative

References

1. Nosratabadi, S.; Mosavi, A.; Shamshirband, S.; Kazimieras Zavadskas, E.; Rakotonirainy, A.; Chau, K.W. Sustainable Business Models: A Review. *Sustainability* **2019**, *11*, 1663. [CrossRef]
2. Boons, F.; Lüdeke-Freund, F. Business models for sustainable innovation: State-of-the-art and steps towards a research agenda. *J. Clean. Prod.* **2013**, *45*, 9–19. [CrossRef]
3. Stubbs, W.; Cocklin, C. Conceptualizing a "Sustainability Business Model". *Organ. Environ.* **2008**, *21*, 103–127. [CrossRef]
4. Dentchev, N.; Rauter, R.; Jóhannsdóttir, L.; Snihur, Y.; Rosano, M.; Baumgartner, R.; Nyberg, T.; Tang, X.; van Hoof, B.; Jonker, J. Embracing the variety of sustainable business models: A prolific field of research and a future research agenda. *J. Clean. Prod.* **2018**, *194*, 695–703. [CrossRef]
5. Lüdeke-Freund, F.; Dembek, K. Sustainable business model research and practice: Emerging field or passing fancy? *J. Clean. Prod.* **2017**, *168*, 1668–1678. [CrossRef]
6. Bocken, N.M.P.; Short, S.W.; Rana, P.; Evans, S. A literature and practice review to develop sustainable business model archetypes. *J. Clean. Prod.* **2014**, *65*, 42–56. [CrossRef]
7. Broman, G.I.; Robert, K.-H. A framework for strategic sustainable development. *J. Clean. Prod.* **2015**, *140*, 1–15. [CrossRef]
8. Evans, S.; Vladimirova, D.; Holgado, M.; Van Fossen, K.; Yang, M.; Silva, E.A.; Barlow, C.Y. Business Model Innovation for Sustainability: Towards a Unified Perspective for Creation of Sustainable Business Models. *Bus. Strategy Environ.* **2017**, *26*, 597–608. [CrossRef]
9. Schaltegger, S.; Hansen, E.G.; Lüdeke-Freund, F. Business Models for Sustainability: Origins, Present Research, and Future Avenues. *Organ. Environ.* **2016**, *29*, 3–10. [CrossRef]
10. Upward, A.; Jones, P.H. An Ontology for Strongly Sustainable Business Models: Defining an Enterprise Framework Compatible with Natural and Social Science. *Organ. Environ.* **2015**, *29*, 97–123. [CrossRef]
11. Vial, G. Understanding digital transformation: A review and a research agenda. *J. Strategy Inf. Syst.* **2019**, *28*, 118–144. [CrossRef]
12. Chesbrough, H. Business model innovation: It's not just about technology anymore. *Strategy Leadersh.* **2007**, *35*, 12–17. [CrossRef]
13. Chesbrough, H. Business model innovation: Opportunities and barriers. *Long Range Plann.* **2010**, *43*, 354–363. [CrossRef]
14. Lüdeke-Freund, F. Towards a Conceptual Framework of Business Models for Sustainability. In Proceedings of the Knowledge Collaboration & Learning for Sustainable Innovation—Conference Proceedings, 14th European Roundtable on Sustainable Consumption and Production (ERSCP) & 6th Environmental Management for Sustainable Universities (EMSU), Delft, The Netherlands, 25–29 October 2010.
15. Schaltegger, S.; Lüdeke-Freund, F.; Hansen, E.G. Business Cases for Sustainability The Role of Business Model Innovation for Corporate Sustainability. *Int. J. Innov. Sustain. Dev.* **2012**, *6*, 95. [CrossRef]
16. Bouwman, H.; Nikou, S.; Molina-Castillo, F.J.; De Reuver, M. The Impact of Digitalization on Business Models. *Digit. Policy Regul. Gov.* **2018**, *20*, 105–124. [CrossRef]
17. Parida, V.; Sjödin, D.; Reim, W. Reviewing Literature on Digitalization, Business Model Innovation, and Sustainable Industry: Past Achievements and Future Promises. *Sustainability* **2019**, *11*, 391. [CrossRef]
18. Rayna, T.; Striukova, L. From rapid prototyping to home fabrication: How 3D printing is changing business model innovation. *Technol. Forecast. Soc. Chang.* **2016**, *102*, 214–224. [CrossRef]
19. Tiscini, R.; Testarmata, S.; Ciaburri, M.; Ferrari, E. The blockchain as a sustainable business model innovation. *Manag. Decis.* **2020**, *58*. [CrossRef]
20. Amit, R.; Zott, C. *Business Model Innovation Strategy: Transformational Concepts and Tools for Entrepreneurial Leaders*; John Wiley & Son Ltd.: Hoboken, NJ, USA, 2020.
21. Melville, N.; Kraemer, K.; Gurbaxani, V.; Carroll, W.E. Review: Information Technology and Organizational Performance: An Integrative Model of IT Business Value. *MIS Q.* **2004**, *28*, 283–322. [CrossRef]
22. Dao, V.; Langella, I.; Carbo, J. From green to sustainability: Information Technology and an integrated sustainability framework. *J. Strategy Inf. Syst.* **2011**, *20*, 63–79. [CrossRef]
23. Melville, N.P. Information Systems Innovation for Environmental Sustainability. *MIS Q.* **2010**, *34*, 1–21. [CrossRef]
24. Watson, R.T.; Boudreau, M.-C.; Chen, A.J. Information Systems and Environmentally Sustainable Development: Energy Informatics and New Directions for the IS Community. *Source MIS Q.* **2010**, *34*, 23–38. [CrossRef]
25. Boudreau, M.-C.; Chen, A.; Huber, M. *Green IS: Building Sustainable Business Practices*; Watson, R.T., Ed.; Information Systems, Global Text Project: Athens, GA, USA, 2008.
26. Gholami, R.; Sulaiman, A.B.; Ramayah, T.; Molla, A. Senior managers' perception on green information systems (IS) adoption and environmental performance: Results from a field survey. *Inf. Manag.* **2013**, *50*, 431–438. [CrossRef]
27. Molla, A.; Cooper, V.; Pittayachawan, S. IT and eco-sustainability: Developing and validating a green IT readiness model. In Proceedings of the Thirtieth International Conference on Information Systems, Phoenix, AZ, USA, 15–18 December 2009.
28. Faucheux, S.; Nicolaï, I. IT for green and green IT: A proposed typology of eco-innovation. *Ecol. Econ.* **2011**, *70*, 2020–2027. [CrossRef]
29. Hedman, J.; Henningsson, S.; Selander, L. Organizational Self-Renewal: The Role of Green Is in Developing Eco-effectiveness. In Proceedings of the ICIS 2012 Proceedings, Orlando, FL, USA, 16–19 December 2012.
30. Molla, A.; Abareshi, A.; Cooper, V. Green IT beliefs and pro-environmental IT practices among IT professionals. *Inf. Technol. People* **2012**, *27*, 129–154. [CrossRef]

31. Kane, G.C.; Palmer, D.; Phillips, A.N.; Kiron, D.; Buckley, N. *Achieving Digital Maturity*; MIT Sloan Management Review; Deloitte University Press: Boston, MA, USA, 2017.
32. Kane, G.C.; Palmer, D.; Phillips, A.N.; Kiron, D.; Buckley, N. *Coming of Age Digitally: Learning, Leadership, and Legacy*; MIT Sloan Management Review; Deloitte Insights: Boston, MA, USA, 2018.
33. Sebastian, I.M.; Mocker, M.; Ross, J.W.; Moloney, K.G.; Beath, C.; Fonstad, N.O. How Big Old Companies Navigate Digital Transformation. *MIS Q. Exec.* **2017**, *16*, 197–213.
34. Hanelt, A.; Busse, S.; Kolbe, L.M. Driving business transformation toward sustainability: Exploring the impact of supporting IS on the performance contribution of eco-innovations. *Inf. Syst. J.* **2017**, *27*, 463–502. [CrossRef]
35. Hildebrandt, B.; Hanelt, A.; Firk, S. Sharing Yet Caring: Mitigating Moral Hazard in Access-Based Consumption through IS-Enabled Value Co-Capturing with Consumers. *Bus. Inf. Syst. Eng.* **2018**, *60*, 227–241. [CrossRef]
36. Kathan, W.; Matzler, K.; Veider, V. The sharing economy: Your business model's friend or foe? *Bus. Horiz.* **2016**, *59*, 663–672. [CrossRef]
37. Gil-Gomez, H.; Guerola-Navarro, V.; Oltra-Badenes, R.; Lozano-Quilis, J.A. Customer relationship management: Digital transformation and sustainable business model innovation. *Econ. Res. Istraž.* **2020**, *33*, 2733–2750. [CrossRef]
38. Ockwell, D.; Atela, J.; Mbeva, K.; Chengo, V.; Byrne, R.; Durrant, R.; Kasprowicz, V.; Ely, A. Can Pay-As-You-Go, Digitally Enabled Business Models Support Sustainability Transformations in Developing Countries? Outstanding Questions and a Theoretical Basis for Future Research. *Sustainability* **2019**, *11*, 2105. [CrossRef]
39. Dumitriu, D.; Militaru, G.; Deselnicu, D.C.; Niculescu, A.; Popescu, M.A.-M. A Perspective Over Modern SMEs: Managing Brand Equity, Growth and Sustainability Through Digital Marketing Tools and Techniques. *Sustainability* **2019**, *11*, 2111. [CrossRef]
40. Jose, C.; Jabbour, C.; De Sousa Jabbour, A.B.L.; Sarkis, J.; Filho, G. Unlocking the circular economy through new business models based on large-scale data: An integrative framework and research agenda. *Technol. Forecast. Soc. Chang.* **2019**, *144*, 546–552. [CrossRef]
41. Brenner, B. Transformative Sustainable Business Models in the Light of the Digital Imperative—A Global Business Economics Perspective. *Sustainability* **2018**, *10*, 4428. [CrossRef]
42. European Commission Communication from the Commission to the European Parliament, the Council, the European Central Bank, the European Economic and Social Committee, the Committee of the Regions and the European Investment Bank Annual Growth Survey. 2018. Available online: https://www.eumonitor.eu/9353000/1/j9vvik7m1c3gyxp/vkjl90nh6wxh (accessed on 31 July 2018).
43. Seele, P.; Lock, I. The game-changing potential of digitalization for sustainability: Possibilities, perils, and pathways. *Sustain. Sci.* **2017**, *12*, 183–185. [CrossRef]
44. Moher, D.; Liberati, A.; Tetzlaff, J.; Altman, D.G. The PRISMA Group Preferred Reporting Items for Systematic Reviews and Meta-Analyses: The PRISMA Statement. *PLoS Med.* **2009**, *6*. [CrossRef]
45. Levy, Y.; Ellis, T. A Systems Approach to Conduct an Effective Literature Review in Support of Information Systems Research. *Inf. Sci.* **2006**, *9*, 181–212. [CrossRef]
46. Kitchenham, B.; Charters, S. *Guidelines for Performing Systematic Literature Reviews in Software Engineering*; Software Engineering Group, School of Computer Science and Mathematics, Keele University: Staffs, UK; Department of Computer Science, University of Durham: Durham, UK, 2007.
47. Elo, S.; Kyngäs, H. The qualitative content analysis process. *J. Adv. Nurs.* **2008**, *62*, 107–115. [CrossRef]
48. Mayring, P. *Qualitative Content Analysis. Theoretical Foundation, Basic Procedures and Software Solution*; Social Science Open Access Repository: Klagenfurt, Austria, 2014.
49. Hsieh, H.F.; Shannon, S.E. Three approaches to qualitative content analysis. *Qual. Health Res.* **2005**, *15*, 1277–1288. [CrossRef]
50. Flick, U. *An Introduction to Qualitative Research*, 4th ed.; Sage Publications Ltd.: New York, NY, USA, 2009.
51. Tableau. Available online: https://www.tableau.com/ (accessed on 16 April 2020).
52. Leximancer. Available online: https://info.leximancer.com/ (accessed on 16 April 2020).
53. Randhawa, K.; Wilden, R.; Hohberger, J. A Bibliometric Review of Open Innovation: Setting a Research Agenda. *J. Prod. Innov. Manag.* **2016**, *33*, 750–772. [CrossRef]
54. Pucihar, A. The digital transformation journey: Content analysis of Electronic Markets articles and Bled eConference proceedings from 2012 to 2019. *Electron. Mark.* **2020**, *30*, 29–37. [CrossRef]
55. Geissdoerfer, M.; Vladimirova, D.; Evans, S. Sustainable business model innovation: A review. *J. Clean. Prod.* **2018**, *198*, 401–416. [CrossRef]
56. Bocken, N.; Smeke Morales, L.; Lehner, M. Sufficiency Business Strategies in the Food Industry—The Case of Oatly. *Sustainability* **2020**, *12*, 824. [CrossRef]
57. Piscicelli, L.; Ludden, G.D.S.; Cooper, T. What makes a sustainable business model successful? An empirical comparison of two peer-to-peer goods-sharing platforms. *J. Clean. Prod.* **2018**, *172*, 4580–4591. [CrossRef]
58. Zott, C.; Amit, R.; Massa, L. The Business Model: Recent Developments and Future Research. *J. Manag.* **2011**, *37*, 1019–1042. [CrossRef]
59. Yang, M.; Evans, S.; Vladimirova, D.; Rana, P. Value uncaptured perspective for sustainable business model innovation. *J. Clean. Prod.* **2017**, *140*, 1794–1804. [CrossRef]
60. Teece, D.J. Business models, business strategy and innovation. *Long Range Plan.* **2010**, *43*, 172–194. [CrossRef]

61. Freudenreich, B.; Lüdeke-Freund, F.; Schaltegger, S. A Stakeholder Theory Perspective on Business Models: Value Creation for Sustainability. *J. Bus. Ethics* **2019**, *166*, 1–16. [CrossRef]
62. Richardson, J. The Business Model: An Integrative Framework for Strategy Execution. *Strateg. Chang.* **2008**, *17*, 133–144. [CrossRef]
63. Osterwalder, A.; Pigneur, Y. *Business Model Generation*; John Wiley and Sons: New York, NY, USA, 2010.
64. Baldassarre, B.; Calabretta, G.; Bocken, N.; Jaskiewicz, T. Bridging sustainable business model innovation and user-driven innovation: A process for sustainable value proposition design. *J. Clean. Prod.* **2017**, *147*, 175–186. [CrossRef]
65. Vladimirova, D. Building Sustainable Value Propositions for Multiple Stakeholders: A Practical Tool. *J. Bus. Model.* **2019**, *7*, 1–8.
66. Jabłonski, M. Value Migration to the Sustainable Business Models of Digital Economy Companies on the Capital Market. *Sustainability* **2018**, *10*, 3113. [CrossRef]
67. Chian Tan, F.T.; Cahalane, M.; Tan, B.; Englert, J. How GoGet CarShare's Product-Service System is Facilitating Collaborative Consumption. *MIS Q. Exec.* **2017**, *16*, 265–277.
68. Heiskala, M.; Jokinen, J.-P.; Tinnilä, M. Crowdsensing-based transportation services—An analysis from business model and sustainability viewpoints. *Res. Transp. Bus. Manag.* **2016**, *18*, 38–48. [CrossRef]
69. Gössling, S.; Hall, M. Sharing versus collaborative economy: How to align ICT developments and the SDGs in tourism? *J. Sustain. Tour.* **2019**, *27*, 74–96. [CrossRef]
70. Breidbach, C.F.; Maglio, P.P. Technology-enabled value co-creation: An empirical analysis of actors, resources, and practices. *Ind. Mark. Manag.* **2016**, *56*, 73–85. [CrossRef]
71. Malhotra, A.; Melville, N.P.; Ross, S.M.; Watson, R.T. Spurring Impactful Research on Information Systems for Environmental Sustainability. *MIS Q.* **2013**, *37*, 1265–1274. [CrossRef]
72. Cooper, V.; Molla, A. Information systems absorptive capacity for environmentally driven IS-enabled transformation. *Inf. Syst. J.* **2017**, *27*, 379–425. [CrossRef]
73. Ripple, W.J.; Wolf, C.; Newsome, T.M.; Galetti, M.; Alamgir, M.; Crist, E.; Mahmoud, M.I.; Laurance, W.F. World Scientists' Warning to Humanity: A Second Notice. *Bioscience* **2017**, *67*, 1026–1028. [CrossRef]
74. Lessambo, F.I. Corporate Governance in Continental Europe. In *The International Corporate Governance System*; Palgrave Macmillan: London, UK, 2014; pp. 114–129.
75. Lessambo, F.I. Corporate Governance in the United States of America. In *The International Corporate Governance System*; Palgrave Macmillan: London, UK, 2014; pp. 46–80.
76. Alt, R. Electronic Markets on sustainability. *Electron. Mark.* **2020**, *30*, 667–674. [CrossRef]
77. Gholami, R.; Watson, R.T.; Hasan, H.; Molla, A.; Bjorn-Andersen, N. Information Systems Solutions for Environmental Sustainability: How Can We Do More? *J. Assoc. Inf. Syst.* **2016**, *17*. [CrossRef]
78. Elliot, S.; Webster, J. Editorial: Special issue on empirical research on information systems addressing the challenges of environmental sustainability: An imperative for urgent action. *Inf. Syst. J.* **2017**, *27*, 367–378. [CrossRef]
79. Jeansson, J.; Bredmar, K. Digital Transformation of SMEs: Capturing Complexity. In *Humanizing Technology for a Sustainable Society, Proceedings of the 32nd Bled eConference, Bled, Slovenia, 16–19 June 2019*; Pucihar, A., Kljajić Borštnar, M., Bons, R., Seitz, J., Cripps, H., Vidmar, D., Eds.; University of Maribor Press: Maribor, Slovenia, 2019; pp. 523–541.
80. European Commission. *Member State Implementation of Directive 2014/95/EU*; European Commission: Brussels, Belgium, 2017.
81. European Parliament and the Council of the European Union. *DIRECTIVE 2014/95/EU*; European Parliament and the Council of the European Union: Strasbourg, France, 2014.
82. Bressanelli, G.; Adrodegari, F.; Perona, M.; Saccani, N. Exploring How Usage-Focused Business Models Enable Circular Economy through Digital Technologies. *Sustainability* **2018**, *10*, 639. [CrossRef]
83. Camacho-Otero, J.; Boks, C.; Nilstad Pettersen, I. Consumption in the Circular Economy: A Literature Review. *Sustainability* **2018**, *10*, 2758. [CrossRef]
84. Seidel, S.; Bharati, P.; Fridgen, G.; Watson, R.T.; Albizri, A.; Boudreau, M.-C.; Butler, T.; Kruse, L.C.; Guzman, I.; Karsten, H.; et al. The Sustainability Imperative in Information Systems Research. *Commun. Assoc. Inf. Syst.* **2017**, *40*, 3. [CrossRef]
85. Parida, V.; Wincent, J. Why and how to compete through sustainability: A review and outline of trends influencing firm and network-level transformation. *Int. Entrep. Manag. J.* **2019**, *15*, 1–19. [CrossRef]
86. Hedman, J.; Henningsson, S. Developing ecological sustainability: A green IS response model. *Inf. Syst. J.* **2016**, *26*, 259–287. [CrossRef]
87. Kurkalova, L.A.; Carter, L. Sustainable production: Using simulation modeling to identify the benefits of green information systems. *Decis. Support Syst.* **2017**, *96*, 83–91. [CrossRef]
88. European Commission. *Monitoring the Digital Economy & Society 2016–2021*; European Commission: Brussels, Belgium, 2015; Volume 52. [CrossRef]
89. European Commission. *Integration of Digital Technology: Europe's Digital Progress Report 2017*; European Commission: Brussels, Belgium, 2017.
90. Levstek, A.; Hovelja, T.; Pucihar, A. IT Governance Mechanisms and Contingency Factors: Towards an Adaptive IT Governance Model. *Organizacija* **2018**, *51*, 286–310. [CrossRef]

MDPI
St. Alban-Anlage 66
4052 Basel
Switzerland
Tel. +41 61 683 77 34
Fax +41 61 302 89 18
www.mdpi.com

Sustainability Editorial Office
E-mail: sustainability@mdpi.com
www.mdpi.com/journal/sustainability

www.ingramcontent.com/pod-product-compliance
Lightning Source LLC
LaVergne TN
LVHW070644170726
843515LV00004B/321

* 9 7 8 3 0 3 6 5 4 3 0 5 5 *